Nuclear Security: Strategies and Techniques

NUCLEAR SECURITY: STRATEGIES AND TECHNIQUES

I.K. KHAN

MD Publications Pvt Ltd
New Delhi

Published by
MD Publications Pvt Ltd
"MD House", 11, Darya Ganj
New Delhi-110002
Tel. : +91-11-41563325, 41562846
E-mail : contact@mdppl.com
Website : www.mdppl.com

ISBN : 978-81-7533-114-3

PRICE : Rs. 1295.00

Published and Printed by **MD Publications Pvt Ltd** at Times Press,
910, Jatwara Street, Darya Ganj, New Delhi.

CONTENTS

Preface

The modern society, whether in developed or in developing countries, depend on the availability of nuclear energy and on the day-to-day use of radioactive materials in medicine, agriculture, industry and for research. Before 9/11, these activities were mainly covered by safety rules regarding health and environment. Since 9/ 11, it is clear, that these activities also require adequate security. For the continued, and expanded, use of nuclear energy or radioactive materials, nuclear security is indispensable and an important prerequisite for successful and sustainable development.

Many of our nuclear security services, expert assistance and training events, we have assisted Member States in their efforts to improve their preparedness and response capabilities and acquired a much better understanding of Member States problems and concerns and the need for further support. The end of the Cold War was marked by a shift from a bi-polar structure of global security into a more complex and unpredictable configuration of world affairs. It also brought about new security challenges, i.e. an increased probability for low-density regional, national or sub-national conflicts with new and more dispersed threats emanating from a larger number of actors, including non-state actors; terrorists or criminals. The audio-visual impact of modern media has dramatically enhanced the socio-psychological impact on a global scale of such conflicts. The number of cases of illicit trafficking in nuclear materials that were recorded since the 90's raised concern about the international physical protection regime and triggered an effort to enhance our capabilities for prevention, detection and responses regarding terrorist acts, as well as to strengthen the Convention on the Physical Protection of Nuclear Material.

Immediately after the events of 9/11, based on the re-evaluation of its implications, the IAEA identified four types of threat for nuclear security: a) theft of a nuclear weapon; b) construction of a crude nuclear explosive device using stolen nuclear materials; c) malicious use of nuclear and other radioactive materials including Radiological Dispersal Devices (RDD); and d) an attack on or sabotage of a nuclear installation or transported materials. The potential targets of such acts include nuclear power plants, fuel cycle facilities, research reactors, laboratories and storages as well as locations all over the world where these substances are used in a broad range of non-nuclear applications.

To prevent these events from happening, we must have a comprehensive, global approach to nuclear security, based on internationally accepted instruments, and which is implemented worldwide and in broad partnerships. Should a nuclear terrorist act happen, we would all suffer, directly or indirectly, as fellow passengers in the same boat.

The expanded use of nuclear energy technology as well as rapidly growing science and medical use of radioisotopes are evidence of the important role of nuclear technologies in sustainable development. The privatisation of the nuclear power industry, deregulation and government reform, point to expanded security related responsibilities for the private sector and other NGOs. Thus, international consensus on the establishing and enhancing of global nuclear security framework is urgently needed.

A strengthened global nuclear security framework requires useful information to have good understanding of the threats, risks, and world-wide status of nuclear security. It also requires effective long-term measures to prevent any terrorist from completing successfully a malicious act, as well as measures to detect and respond to smuggling or theft of nuclear materials or radioactive substances. There is also a need to continue efforts to reduce the threat by eliminating, as much and as quickly as possible, the quantities of highly enriched uranium or plutonium from peaceful applications for which they are not needed. Finally, it requires measures to improve security of poorly protected nuclear installations and transport of nuclear and radioactive materials.

Thanks to Mr. Pranav Gupta of **MD Publications Pvt Ltd**, New Delhi, who supported me fully for completing this book.

I.K. KHAN

1

Nuclear Security: Technology and Experience

Introduction

A very limited and frankly myopic -view of the relationship between technology and security equates the mere quantity or sophistication of the former with the (presumed) enhancement of the latter. How tempting the notion that we will somehow be "home free" if we just can get ourselves around the next technological corner! The fact is, however, that no technology or cluster of technologies can alone ensure genuine security, national or international, in this nuclear age.

Eleven days after the bombing of Hiroshima, Secretary of War Henry L. Stimson, together with nuclear physicists J. Robert Oppen heimer, Arthur H.Compton, Enrico Fernit and Ernest O. Lawrence, made the following plea:

We have been unable to devise or propose effective military countermeasures for atomic weapons. Although we realize that future work may reveal possibilities at present obscure to us, it is our firm opinion that no military countermeasures will be found which will be adequately effective in preventing the delivery of atomic weapons. We believe that the safety of this nation as opposed to its ability to inflict damage on an enemy power cannot be wholly or even primarily in its technical prowess. It can only be based on making future wars impossible.

Forty-five years have not diminished the force of this remarkable plea. Indeed, the futility of a guaranteed technological defense against nuclear weapons has been proven repeatedly through almost a half-century of military science. Despite the relentless pursuit of virtually every military technology feasible since World War II, especially on the part of the superpowers, the world the superpowers included is now far less secure than it was when it entered seriously into the arms race in the early 1940s. This is not to say that technology can in no way contribute to the security of a post-nuclear world that, though having abandoned nuclear weapons, is nonetheless capable of "reinventing" them virtually at will. Rather, it is to recognize that, in such a world, new security technologies are at most only a part of the solution, and probably not a very large part at that. It is to recognize that the link between technology and genuine security military, environmental, economic, and so forth -has more to do with the process by which we make technology decisions relating to security than it does with the invention of new and fancier technological hardware.

Application of Technology to Security

An early scene in Stanley Kubrick's film 2001: A Space Odyssey depicts the moment at which an ancestral hominid first conceived of, and used, a rudimentary weapon: a femoral bone with which a member of the same species is bludgeoned to death over competition for food and water. The bone is subsequently thrown with exultation high into the air and thereafter left to tumble to Earth, whereupon it melds into the image of an advanced spacecraft en route to the moon. In this brief moment of cinematic genius, we see the intertwining of weaponry and technology through the millennia to the present, the failure of restraint in the exercise of power expanded artificially beyond the limits set by evolution, the intoxication evinced by its savage application, and the oblivious disregard for future consequences. In this apocryphal scene, we witness the uniquely human conflict between instinct and intellect that has brought humanity to the brink of self-extinction.

Kubrick's triumphant hominids doubtless assumed that, through their new-found weapons technology, they would enjoy a decided advantage indefinitely. But such assumptions never have been borne out in practice and probably never will be. the opposition is human too and, like it or not, has generally the same potential to discover and apply new weapons technologies an annoying detail that typically is dismissed via a repertoire of self-deceptions that define the opposition as intellectually, culturally, and socially inferior, as sub-human. Moreover, breakthroughs that come about more by chance than in response to intellectual effort have the same probability of benefiting one adversary as another. But again we are inclined to deceive ourselves. God favors us, we say, because our cause is just. He can be relied upon, we insist, to discriminate in our favor in the meting out of chance discoveries.

Such rationalizations, though an indispensable part of the arms race, seem silly when written down. And indeed they are! For were they true, and were it so that lasting security could be achieved by technological means, some generation since Kubrick 's hominids would have long ago gotten a permanent leg up on the opposition and held it there. Plainly, lasting security never has been and never will be achieved through technology alone. The other side will learn soon enough how to use a discarded bone as a club and can be expected to improve upon the idea as well. Technological challenges tend to have technological solutions and technological solutions tend to be found by technological beings. Only by debasing the opposition and by invoking Divine Providence can the illusion of security be maintained. It is a tribute to zthe power of ideology (or myth) over reason that otherwise supremely rational beings can cling so tenaciously to gut instincts that fly in the face of the historical evidence. Nothing in the history of human affairs supports the assumption that the balance of nuclear terror or any other military standoff can be maintained indefinitely, the conduct of the arms race since 1945 notwithstanding.

The point especially to be emphasized, however, is that a world free of nuclear weapons, in contradistinction to a completely Strategic bombers built to deliver weapons of mass destruction have so far been confined essentially to mutually exclusive air spaces. The situation is quite different relative to missile-outfitted submarines that navigate in a shared operational area of over 40 million Square miles, the better part of the world's accessible ocean surface.

The problem is likely to be exacerbated in the high technology frontier of space if SDI ever is deployed. Here the laws of nature endow the concept of space-based defense with some rather unwelcome and, as far as one can see, intractable problems and characteristics. One of the most serious is rooted in the very nature of earth orbit itself. Everything is in a state of continuous motion on a global, scale and everything comes close to everything else sooner or later. Just as serious is the fact that space is, by its nature, empty; there is nothing to hide behind or within and nothing to use as camouflage. Space-based assets are therefore inherently vulnerable in the extreme. Virtually all proposed space-based SDI components, across the entire spectrum of technologies, from kinetic-kill weapons to exotic laser systems, and whether for the near or long term, suffer from the profound defect that they are vulnerable to comparatively crude hence cost-effective countermeasures such as orbiting "space mines." The current vogue among some SDI proponents is to suggest, in response to this problem, the maintenance of "keep-out zones" around critical orbiting assets, which is a little like trying to prevent the rain from falling through an eight-foot hole in your roof by declaring the inside of your house a keep-out zone. Nature does not oblige. True, some cling to the belief that technology can always find a way out of any dilemma. But, characteristically, each proposed Band-Aid leads directly to another insuperable problem.

The difficulty of maintaining keep-out zones is compounded by the presence of tens of thousands of objects of varying sizes in orbit around the earth. None have the fuel and control systems that would be required to keep them out of specified zones, and the task of monitoring them is staggering. At best, all objects launched after a certain date could be required, assuming the assent of all States with access to space, to maintain stations outside the forbidden zones. But this leads immediately to another problem: Sooner or later all such station-keeping systems must fail, leaving the associated objects free to drift into the keep-out zones. Though some have recognized that major breakthroughs may be required if the SDI is ever to be feasible, it is hardly reasonable that this should be understood to include the development of space systems with infinite lifetimes. Recognition of this dubious requirement has spawned the proposal that, instead of all other objects in space avoiding the vicinity of the critical assets, the critical assets themselves will be required to avoid all other objects. But this proposal leads directly to yet another absurdity. It requires that the objects to be protected be capable, on their own, of moving away from other objects that approach too closely. Since the SDI components in question would be among the most massive in space, there is no plausible way that the approach of a relatively fast-moving space mine or torpedo could be avoided by such a "step aside" maneuver. Newton's law of inertia cannot be revoked. The idea of automatically destroying anything that drifts into a keep-out zone is equally absurd, even if it were feasible. Imagine the sky being swept clean of thousands of expensive commercial and research satellites, to say nothing of hapless astronauts adrift due to a malfunction, in the context of an automated system that can allow no external override for fear that the opposition might discover how to turn it to its own advantage. And so it goes.

The common thread that binds this second class of problems is a finite environment permeated by ever more sophisticated weapons. Inevitably the protagonists rind themselves defending against each other from within the same garrison, the mutual garrison phenomenon. It is a quirk of locally Euclidian geometry, which will not yield for "man nor beast," that brings with it a plethora of insuperable problems of the most fundamental kind.

Finally, a phenomenon somewhat less general than those considered so far but no less inescapable and with comparable potential to frustrate the goal of security through technological means is derived from a principle I call "the affinity of complexity and obsolescence," applicable primarily to very large integrated technological systems. A prominent characteristic of successive arms race proposals is that they tend toward ever more expansive distributed systems; greater and greater numbers and varieties of components and subsystems are required to realize the overall system mission. this trend is a natural and unavoidable consequence of the parallel trend, already noted, toward progressively more automated globally comprehensive systems of which SDI is so far the ultimate example. As system complexity increases, the problem of systems integration becomes correspondingly more difficult. With the advent of SDI we have reached the point where the difficulty of the integration problem begins to dominate an other technological problems, acute though they may be. The conundrum that attaches to the "affinity of complexity and obsolescence" principle transcends even the system integration problem and lies in waiting should defense managers ever manage successfully to navigate the integration waters.

As system complexity increases, it becomes necessary earlier and earlier in the development phase to commit to technologies that will be used in the completed system. This is unavoidable because the ramifications associated with changes of technology become more and more difficult to isolate and deal with as overall complexity increases. And, of course, added difficulty translates into added time, so that the penalty for the accommodation of newer technologies in midstream is, generally, postponement of the project completion date and escalating costs by increments that increase with complexity. There emerges a tradeoff between the ability to introduce newer technological developments in midstream and the ability to deliver the final operational system. The less complex the system, the more readily changes can be accommodated and the more closely the final product can reflect the technological state of the art. Conversely, the more complex the system, the less readily can changes be accommodated, with the result that the technologies embodied in the final product tend more and more toward obsolescence. Hence: the affinity of complexity and obsolescence.

All of which bodes ill for anyone still taking SDI or comparable schemes seriously. Considering its unprecedented complexity, it may well have to be deployed with technology twenty-five or more years out of date. And, after initial deployment, the extraordinary difficulty of retrofitting modifications, both in terms of cost and of assuring the maintenance of reliable systems integration, will bias SDI toward even greater obsolescence. Extremely complex distributed systems such as SDI are born with an exposed Achilles tendon, whereas effective countermeasures, being required only to foil the opponent's systems by any means (however lacking in sophistication), enjoy the advantages of reliability and cost effectiveness that attend considerably reduced complexity. Ironically, the hope for security through technological means becomes less and less realistic as systems become more and more comprehensive and complex.

Thus we see, in the preceding paragraphs, some of the intrinsic factors that limit humankind's ability to achieve security indefinitely through technological means. It should be noted that none of these factors derive from the particulars of the systems or techniques involved; they are intrinsic, arising independently of specific concepts, and have become evident only recently due to the level of technological complexity and sophistication that, at the close of the Twentieth Century, we finally have been able to achieve. Though these limits are an anathema to persons bound by traditional ideas of what ought to be possible, their fundamental character and foreboding implications can be ignored only at our peril.

Helpful Technological Options

The achievement and maintenance of global security exclusively through technological means, technology can make at least a few, possibly even some indispensable, contributions to the achievement and maintenance of an essentially nuclear-weapons-free world. Just as technology has contributed to the arms race the raison d'être of which resides, at bottom, with technology so also might technology contribute to the peace race. In particular, technology can be helpful insofar as it assists the world community and its constituent polities to become more anticipatory and preventive rather than merely reactive in the execution of security strategy. The menace embodied in nuclearism is of such character that technological and other security measures that fail to anticipate and prevent are not especially valuable. Thus, technologies that focus on the verification and enforcement of nuclear arms reductions and on the monitoring of relative differences in conventional forces would appear promising. Of course, other as yet unknown technologies surely will emerge in the years to come to assist the promotion and protection of a post-nuclear world.

While we consider suggestions for technologies that could assist a post-nuclear global security system, however, it must be borne in mind that the very idea of such technologies makes sense only in the context of an overriding geopolitical regime that is dedicated to the permanent end of arms racing. There simply is no hope of lasting and meaningful security through technological means alone. Indeed, as discussed above, it is advanced technology itself that has driven us to the point that any margin of safety temporarily gained is inevitably eroded and then followed by even greater danger than existed at the outset. This is of course precisely the position in which the world finds itself after forty-five years of attempting security through ever more advanced technological proposals and "solutions."

Also, it is important to proceed cautiously in the definition and application of alternative security technologies, especially in the wake of the recent sweeping reforms in the Soviet Union and Eastern Europe and the parallel emergence of strange national and international security bedfellows. Arms controllers and disarmers, formerly critical of defense contractors, now offer helpful defense suggestions; and defense contractors, accustomed to looking upon arms controllers and disarmers with jaundiced eyes, now listen to these suggestions both cordially and attentively. But unless an irreversible shift really does take place, parties with traditionally conflicting goals can become too easily the unwitting servants of the other. In the absence of a fundamental change in international relations, defense contractors could be entering merely into a kind of holding pattern, awaiting the return of a cold or even hot war in the competition for scarce resources or the battle against drugs, for example and in the process pining the acquiescence, even the active cooperation, of their former critics.

Now, with these preliminary caveats in mind, let us turn to those anticipatory and preventive technologies that seem the most promising for a post-nuclear world, bearing in mind that among my present purposes, as shall be seen, is an urgent desire to provoke engineers, physicists, and other technical scientists to reconsider as much as possible their relationship, if any, to the military industrial complex.

Surveillance Satellites

A set of technologies especially well suited to anticipating and preventing resort to force, even small-scale uses of force, is that of orbiting satellites and remote sensors. Both in the East and West, surveillance satellites currently carry a wide array of sensors - photographic cameras, multispectral scanners, infrared sensors, microwave radars, and electronic listening devices that are capable of obtaining many different types of information from the earth's surface; and while their orbital paths, their operating wavelengths, and the intervening atmosphere impose certain fundamental limits on their effectiveness, they are becoming both increasingly sensitive and increasingly precise.

As part of a complex of safeguards designed to ensure a world free of nuclear weapons, surveillance satellites, available to national and international security managers, could perform a number of critical functions. Most obviously, they could help to guarantee compliance with arms control and arms reduction agreements, exposing violations when they occur, which then could be reported and discussed. They even could be used to spot the deployment of chemical weapons and factories capable of making them. Similarly, they could be used to monitor troop movements and conventional force deployments so that potential conflicts might be addressed at an earlier stage with the goal of preventing an escalation of crisis. Finally, they could be used to facilitate timely and convenient intergovernmental communication.

The French government proposed to the United Nations General Assembly the establishment of an international satellite monitoring agency that would oversee most if not all of the potential functions of satellite surveillance mentioned above. Such an agency would include an international team of experts that could analyze, interpret, and disseminate retrieved security information to the member nations and, by so doing, engender increased confidence in the evolving world system. And, because such an enterprise would require major cooperation and exchange, it would likely simultaneously enhance international communication and understanding, which, in turn, could further contribute to increased international stability

Seismological Verification

An international network of seismological stations could help to verify compliance with nuclear test bans, comprehensive and otherwise. In 1976 in Geneva, the United Nations Committee on Disarmament proposed a set of guidelines that could be used to implement a program of cooperation among national seismological facilities. In such a system, which would benefit nations that do not have the resources to accomplish the task, central offices would interpret the compiled data and prepare comprehensive reports so that judgments about the causes of different disturbances could then be made with confidence and in unanimity. Combined with satellite verification, such a network could make it extremely difficult to test nuclear weapons without alerting the international community.

Radar Systems

In a nuclear-weapons-free world, radar systems, like surveillance satellites, could be used to assist the goals of anticipation and prevention. International peacekeeping forces as well as national military establishments could benefit from such technology by obtaining early warning of potential aggression. Examples of modern developments in this field include sophisticated surveillance systems mounted on reconnaissance aircraft, typified by the US Joint Surveillance Target Attack Radar System (JUSTARS), and over-the-horizon radar (OTH), which is diffracted and scattered by the earth's atmosphere to achieve greater range than conventional systems.

Submarine-Tracking Technology

Technologies capable of locating and tracking submarines also can contribute to the overall security of a nuclear-weapons-free world. In addition to sonar, a number of technologies have been devised or are under investigation for this purpose, including acoustical surveillance, the monitoring of heat emissions, the use of lasers, the detection of a submarine's magnetic field, and the observation of surface or interior waves. Although the feasibility of a reliable submarine detection system is a matter of scientific dispute, the development of such a system could be another step in providing security against submarines that carry conventional weapons or, in the case of treaty violations, nuclear weapons.

High-Tech Conventional Defense

Even in a radically altered global system, free from the current nuclear standoff, the problem of disparities in conventional armed force structures will persist. The development of sophisticated defensive weapons, however, in the form of, for example, remotely piloted vehicles (RPVs) and precision-guided munitions (PGMs) i.e., accurate, target-seeking weapons launched from ground-based artillery systems or aircraft could help to reduce or even eliminate the consequences of such imbalances. The potential advantage of such weapons and weapons systems is that, when integrated into an intelligent defensive scheme, they make conventional invasion difficult or impossible and are, in addition, far less expensive than the weapons they are designed to destroy, i.e., armored vehicles, aircraft, and surface ships. Several plans of this type already have been suggested for the conventional defense of Western Europe.

Security Depends on More than Technology

The ultimate utility of the foregoing and like technologies in helping to safeguard a post-nuclear world necessarily will depend on basic changes in other dimensions of the global system. Without corresponding reforms or transformations in the socioeconomic, political legal and other realms, technological answers to the national and international security problem are more likely to perpetuate rather than assist the struggle for progressive change. Without an overriding geopolitical regime whose endpoint is the termination of arms-racing once and for all, the selfsame technologies could prove catastrophically destabilizing. For example, stability enhanced by mutual satellite surveillance could turn into its opposite in a time of resumed cold war crisis. Under such circumstances, satellite observations otherwise satisfactorily explicable or not seen at all with lesser technology could be interpreted as threatening, leading to responses that likewise could be interpreted as threatening, thus creating an ominous self-fulfilling run-away situation compounding rather than allaying fear and suspicion of imminent danger. Remotely piloted vehicles and precision-guided munitions, while holding the line in times of trust, can as well be used with devastating consequence in times of war. One has but to reflect on the role of the machine gun or the introduction of aircraft into World War I to appreciate the downside of nearly every alternative security technology.

Also, it bears notice that the technologies briefly noted here derive from a definition of security that emphasizes the avoidance of military confrontation and conflict. A broader conception, informed by a desire to avoid mounting ecological and economic threats, would result, of course, in a wider array of technological recommendations aimed at thwarting environmental decay, stimulating economic well-being, and otherwise enhancing the human condition. Form follows function.

But considerations such as these are beyond the primary purposes of this chapter, which, in addition to recommending some technological options to help safeguard a post-nuclear world, are to demonstrate the inherent limits of technology relative to security and to expose the highly questionable ways in which technology-security decisions have heretofore been made, with a view to suggesting an alternative to the nuclear threat system upon which, to our enormous insecurity, we currently rely. It is my fundamental belief that no amount of technological innovation least of all such assumed marvels as the Strategic Defense Initiative or SDI, which I prefer to call the "Self-Deception Initiative" can substitute for informed, humane, and rational decision-making in the quest for lasting security. There exists a great tendency to engage the national and international security debate at the level of specific weapons and technologies. Interesting as such debates may be, however, they forestall serious consideration of the ways in which the application of technology to security has been perverted and, as well, the ways in which technology can be made to serve the fundamental socioeconomic and political bases of true security so as to benefit all humankind.

Misapplication of Technology to Security

It is a timeless maxim that we can use technology either to promote the common good or to pursue shortsighted, self-serving interests. Technology itself has no morality; it is only the applications of technology, arising as they do from human choices, that can be judged by moral standards. Technology, and the opportunities it affords, do not arise spontaneously out of a vacuum. Humans create technology, and, generally, other humans decide its application. An implicit covenant exists, therefore, between those who create technology and those who apply it.

Unfortunately, humankind has done violence to this covenant, especially when it comes to applying technology to security. Instead of promoting the common good, we too often have misapplied or squandered the precious opportunities that technological advances have made available to us. The history of US national security policy at Hiroshima and Nagasaki during World War II and its conduct in the arms race in the years since demonstrate a generally lamentable quality of decision-making by human beings faced with choices laid open by technology.

Hiroshima-Nagasaki

The US decision to use atomic weapons against Japan in 1945 may be said to have been among the most seriously flawed and regrettable decisions ever made in human history. President Franklin D. Roosevelt approved the Manhattan Project in 1942 to develop an atomic bomb in response to rumors that Nazi Germany already had such a project under way. That is, the US effort was conceived as an attempt to meet a perceived German threat (although, as it turned out, the Germans had decided to "scuttle the project to develop an atom bomb" and instead opted to develop "an energy-producing uranium motor for propelling machinery"). Relative to Japan, however, the facts suggest that far less honorable and altruistic motives were involved.

In the first place, no claim ever has been made that Japan was developing atomic weapons also. Instead, US officials and others have justified the Hiroshima-Nagasaki bombings primarily on the grounds that upwards of 500,000 to 1 million US deaths were averted. In his memoirs, President Harry S. Truman claimed that the bombings saved perhaps 500,000 US combat lives. Secretary of War Henry L. Stimson similarly conjectured "over a million casualties, to American forces alone." And British Prime Minister Winston Churchill claimed in his memoirs that a failure to use atomic weapons "might well require the loss of a million American lives and half the number of British." Most Americans have accepted this "myth information" without question. But credible conflicting evidence strongly suggests that the figures used were more self-serving then real. In mid-June 1945, the US Joint Chiefs of Staff had prepared a plan for the invasion of the southern Japanese island of Kyushu; and, contrary to the figures later popularized by Truman, Stimson, and Churchill, their worst-case estimate of US deaths from the planned invasion was not more than 20,000 and probably less than 15,000 (an estimate that presumed 5,000 air and naval losses prior to the invasion, no more than 10,000 during the invasion itself, and 5,000 unforeseen collateral losses). Indeed, it is difficult to imagine that the Joint Chiefs of Staff would ever have recommended the Japanese invasion plan to the President had the estimates actually been anything like those quoted after the fact.

Also noteworthy is the fact that almost a full month before the fateful Enola Gay flew over Hiroshima, the Japanese were making aggressive peace overtures through the Soviet Union, which had entered the war against Japan at the urgings of Great Britain and the United States at Yalta in February 1945 and had subsequently amassed troops in Manchuria for the purpose of taking over South Sakhalin and the northern Japanese Kurile Islands, as secretly agreed to at Yalta. As early as July 12, 1945, four days prior to the successful Trinity test at Alamogordo, New Mexico, the Japanese Emperor had requested Foreign Minister Shigenori Togo to instruct Japanese Ambassador Naotake Sato in Moscow to inform Soviet Foreign Minister Vyacheslav M. Molotov "that the Emperor wanted the war ended immediately and wished to send Prince Fuminare Konoye to Moscow with power to negotiate a peace on almost any terms, presumably short of the unacceptable sacrifice of the imperial dynasty." These facts were known to the United States on July 13 as a result of intercepted and decoded cables and were carried by Secretary of the Navy James V. Forrestal to President Truman and Secretary of State James F. Byrnes, then meeting with Churchill and Soviet Premier Joseph Stalin at Potsdam. The gravity of the Japanese situation and, hence, the credibility of the surrender overtures are suggested by the precipitous decrease of US aircraft losses from bombing missions over Japan, from a high of 5.7 percent in January 1945 to 0.4 percent by July. Because Japanese air defenses had been so compromised, the United States, by the end of May, had bombed virtually every major city in Japan except, be it noted, Hiroshima and Nagasaki, which, in spite of their accessibility to US bombers, were spared conventional bombing. On the night of March 9 alone, more than 100,000 men, women and children were killed in the incendiary fire-bombing of Tokyo.

Finally, it is important to note that the bomb detonated over Hiroshima on August 6, 1945, producing 68,000 immediate fatalities and 76,000 injured, was an untested gun-type bomb dropped only twenty days after the first successful test of an atomic implosiontype bomb at Alamogordo, New Mexico, on July 16. The Japanese peace overtures, which began prior to the successful Trinity test and which were known to the United States well before August 6, were of no avail; likewise the voices that were raised in a petition to the President circulated by Leo Szilard among his fellow atomic scientists in July 1945, foreswearing the use of atomic weapons against innocent non-combatants. These efforts fell on deaf ears, as did also proposals to demonstrate the power of the bomb before Japanese civilian and military officials prior to its actual use. The stumbling block was President Truman's insistence, in the face of considered recommendations to the contrary from his own staff, on unconditional surrender that is, conditions prohibiting retention of the revered Japanese emperor. In the end, after 38,000 more immediate fatalities and 21,000 more injured at Nagasaki, the Emperor was retained. A weapon conceived in response to a technologically superior German war machine was, instead, used against essentially defenseless, non-combatant Japanese civilians.

It is evident, then, that the rush to use atomic weapons against Japan had less to do with legitimate US military objectives than it did with denying the Soviet Union access to that country. Even then, fearing lost opportunity and despite ongoing attempts on the part of the Japanese to obtain Soviet mediation of terms of surrender, 1.6 million of Stalin's troops attacked Japan from their Manchurian encampments on August 9. The 106,000 dead and 97,000 injured in the opening salvo of the nuclear era thus should be tabulated as the first victims of the Cold War between the United States and the Soviet Union. Never in the history of modern warfare had a major new weapon been taken from development into combat in a matter of a few days, let alone one not tested before combat use, so confident were its designers that it would work on the first try. Disquieting also is the sparing of the cities of Hiroshima and Nagasaki from conventional bombing in anticipation of the atomic bomb, raising the specter of an experiment to determine the lethality and specific effects of the new weapons in a pristine setting uncompromised by prior weapons damage. If, as some have argued, unacceptable casualties had to be inflicted upon civilians to force an early surrender, the evidence from Tokyo makes it clear that the same effect could have been accomplished without resorting to atomic weapons.

In light of the evidence, the use of atomic weapons against Japan was a reprehensible and cowardly act. The first use of atomic weapons by the United States represented what Manhattan Project physicist Leo Szilard referred to as "a flagrant violation of our own moral standards." It may be said also to have constituted a violation of international law and, perhaps most importantly, a violation of the implicit trust between political leadership and the US people, including those whose genius actually created the weapons. But the saga of the arms race is rife with examples of faulty logic and dubious reasoning to justify the misappropriation of technology in the name of national security.

Arms Control

In the field of arms control, for example, the United States has seen itself as committed in principle to the abolition of a system in which (presumed) security depends so heavily on technologically sophisticated weaponry, but has demonstrated itself to be perpetually short of the will to take the first really significant steps mandated by that principle. Indeed, driven by fear verging at times on hysteria and by such other forces as economic greed and the lust for power, the United States has come to measure security in megatons and, consequently, to countenance a long series of negotiated settlements that, by design, have left humanity more imperiled than when the process was entered into in the first place. Whether or not the current reconstitution of Eastern Europe will reverse or retard this practice remains to be seen. But if past behavior is any indication of future conduct we should not be overly optimistic. History shows that "bargaining chips" created by US weapons technology never have been used as such that is, weapons have never been discarded in exchange for a significant arms control objective.

A remarkable opportunity to end the arms race once and for all, largely unknown to the US public, presented itself in 1955 when the United States enjoyed an overwhelming strategic superiority over the Soviet Union and consequently could deal with the Soviets from the much-coveted "position of strength." In December 1954, a joint British-French proposal was put before the United Nations

Disarmament Commission calling for a treaty that would lead to complete nuclear disarmament under international control. The proposal was endorsed by all the Western States, including the United States, with the Soviet Union being the only holdout. In Spring 1955, however, in a move that is not without precedent, the Soviets presented their own proposal incorporating all the main features of the original British-French plan, and the United States responded favorably. Finally, the nuclear powers had agreed upon a formula to end the arms race. The arms race was to be called off.

But the air of optimism and elation was short lived. On September 6, in a dramatic reversal of the US position, the new US delegate to the Commission, Harold Stassen, announced that the United States "put a reservation" on all of the substantive positions it had previously taken in the disarmament commission or at the UN on questions relating to levels of armament. Though then President Dwight D. Eisenhower represented himself publicly as eager to reduce the threat of nuclear war, it is evident that he was caught off guard by the unexpected Soviet acceptance of the terms and conditions officially endorsed by the United States. Embarrassing as the sudden reversal must have been, it was preferable to suffer the temporary criticism that would follow than to be drawn into a policy that would force the United States eventually to abandon its nuclear supremacy. That supremacy, history has shown, was to be eroded eventually, precisely because the arms race was allowed to continue. However, to have seen this possibility and acted accordingly would have required a modicum of vision and statesmanship not generally forthcoming from individuals whose political survival is wedded to the cultivation of the fear of an external enemy. The initial US endorsement was, evidently, disingenuous, and one of the greatest dividends ever yielded on our technological investments was thrown recklessly aside.

Another example of the misappropriation of weapons technology to national security began in 1970 under the Nixon Administration. It has had repercussions to this day, leading, as I will explain, to the "Star Wars" proposal of President Reagan.

By Spring 1970, US weapons technology had advanced to the point of developing the capability of arming individual strategic missiles (launchers) with multiple independently targetable reentry vehicles (MIRVs). Prior to that time, US and Soviet land-based strategic forces formed a stable mutual deterrent because the missiles on each side carried only one reentry vehicle (warhead) per missile a situation of mutual stability that prevailed for many years, with neither side motivated to alter the basic configuration of its land-based forces. The MIRV option, however, was inherently destabilizing. By launching several warheads from each missile, one side could hope to destroy all of the other's launchers and still have missiles to spare for other contingencies. When it became known that the United States was moving to deploy its new MIRV technology on its land-based missiles, the Soviet Union, which did not then have this technology, objected strenuously on the grounds that stability would be lost and the risk of nuclear war thereby increased. At the SALT talks in Vienna in May 1970, the Soviet Union made twenty-six separate overtures to the United States to defer deployment and to enter instead into negotiations to ban MIRV deployment by either side altogether. The response of the Nixon Administration was to refuse to regard these overtures as serious, saying that it was "a trick designed to dupe the United States into delaying MIRV deployment." In characteristic fashion the new technology, developed as a bargaining chip to get the Soviet Union to the negotiating table, was deployed; the bargaining chip was never used as such and a critical opportunity was lost. Having been rebuffed, the Soviets proceeded to develop their own MIRV capability in about five years' time.

But this was not the end of the story. Looking down the Soviet MIRV barrel the United States saw a "window of vulnerability" for its own land-based strategic missiles. Failing to recognize that the problem was of its own making and failing to learn from past mistakes, the United States looked for yet another "technological fix," which soon came in the form of the MX missile proposal. To be shuttled in mobile protective shelters (MPS) along thousands of miles of railroad tracks, the MX was designed to slam shut the alleged window of vulnerability on the US side. When, however, it was recognized that the MX would be the largest human project ever, that it would be extremely expensive, that it would create a national shortage of concrete and have major negative ecological effects, and that, despite its MPS strategy, it still would be a prime target for Soviet missiles, the selling of the project began to bog down.

Enter the Scowcroft Commission, set up by President Reagan to save the MX missile proposal (and named after its chairman, Lieutenant General Brent Scowcroft, currently National Security Advisor to President Bush). Using arguments put forth in vain for years by analysts out of phase with Washington's needs, the Commission came to the conclusion that the window of vulnerability never existed, that the MPS idea should be scrapped, and that the MX missiles should be dropped into existing Minuteman missile silos. The desired effect was achieved: The day was saved for the MX missile and Congress dutifully appropriated the requisite funding. No one asked whether the very concept of land-basing strategic missiles might have been made obsolete by the US and Soviet MIRV decisions. No one asked whether the United States might get a leg up on the Soviet Union simply by abandoning land based missiles in favor of relatively invulnerable strategic submarines and bombers. No one asked whether the Soviets, who had just completed a costly land-based missile modernization program and who had 73 percent of their strategic warheads on land compared to 21 percent for the United States, might be in a most unenviable position. And certainly no one seemed to notice or object to the circular logic involved: The MX proposal was inspired by an alleged window of vulnerability, yet when that window was declared nonexistent by credible authorities, the MX proposal was not consigned to the dust bin.

Of course, unprotected human beings could not be made invulnerable by the fiat of a presidential commission, and so President Reagan proposed to create, in the form of SDI, an impenetrable shield over the United States to protect the US population against nuclear missile attack. The Office of Technology Assessment and all but a cadre of diehards quickly recognized that a general population protection scheme of this sort was not possible for the foreseeable future and probably never would be. But rather than abandon the idea as a pipe dream fraught with major technological problems, Washington took a fall-back position: If SDI would be too leaky to protect human populations, at least it could provide a partial defense for our land-based missiles, thereby enhancing their deterrent value. In this context, naturally, the Scowcroft Commission's finding that no window of vulnerability existed was not to be mentioned too loudly. A number of other observations were also downplayed, such as the possible obsolescence of land-based strategic missiles and the commensurate desirability of abandoning allegedly vulnerable land-based strategic missiles in favor of relatively invulnerable strategic submarines and bombers. SDI was to be deployed to protect land-based missiles, including the MX, which itself was deployed to close an alleged window of vulnerability opened by the earlier failure of the United States to relinquish deployment of MIRV technology. Put another way, if MIRV had not been deployed and instead had been banned by both sides, there would be no more justification for "Star Wars" today than there was at any time after the first installation of land-based missiles and prior to the MIRV decision in 1970. Once again, near-term political ambitions prevailed over long-term considerations of national and international security.

Such circumscribed thinking appears to continue even in the face of a world changed almost beyond recognition as a result of the sweeping reforms introduced by Mikhail Gorbachev, reforms not dreamed of since V.I. Lenin, and now spreading throughout Eastern Europe. Concessions that for the entire course of the arms race have been "pie in the sky" are being granted unilaterally, and proposals that go beyond even our wildest imaginations are being tendered seriously. The Soviets have taken the moral high ground by dint of the US failure to seize prior opportunities opened by its consistently superior strategic technology. Yet, in spite of it an, every major weapons system put into the pipeline in the first Reagan Administration the resurrected B-1 bomber, the Trident II missile, the MX missile, SDI, massive procurements of cruise missiles, the 600 ship Navy, 17,000 new nuclear warheads, the advanced technology (Stealth) bomber and Stealth cruise missile, a massive commitment to C3I (command, control communications, and intelligence) is still in the pipeline or reaching deployment.

Of course, a major explanation for this craziness is to be found in the cast of characters participating in the US application of technology to security, which extends far beyond the politicians and bureaucrats reported daily by the print and broadcast media. Defense contractors and government laboratories, whose interests do not necessarily comport with the greater good, are major players in the arms race, and they enjoy altogether too much influence in the process.

This point is made abundantly clear by the history of the decision to manufacture and deploy the Pershing II missile. Under development by the Martin Marietta Corporation since 1974 as a modernization of the Pershing IA already deployed in Europe, it was to have a more accurate upgraded warhead but otherwise the same short range as the Pershing IA, precluding targeting the Soviet Union from Western Europe. However, early in 1978 Martin Marietta officials became aware that long-range weapons were under discussion in the secret meetings of a special nuclear committee of the NATO ministers known as the "High Level Group" (HLG). Moving quickly, Martin Marietta proposed directly to US Army officials and members of the HLG that a second-stage rocket be added to the missile as part of its upgrade, enabling it easily to strike targets within the Soviet Union. The marketing maneuver succeeded. Persuaded that the addition of a second-stage rocket would be largely overlooked in what otherwise would be perceived as a routine upgrade, the NATO ministers accepted the Martin Marietta proposal and left it to the United States to determine the precise number of missiles to be deployed and the types of warheads they were to carry. Stated Hans Apel, then the West German defense minister, "we didn't anticipate any problems. It looked so simple just to replace the Pershing IAs with more modern weapons. Nobody thought about any political repercussions." Though promoted subsequently in response to the deployment of the Soviet SS-20 missile (itself a modernization of the aging Soviet SS-4 and SS-5 missiles), the Pershing II that finally emerged was more the result of a marketing ploy than a considered response to a perceived military need an application of technology to security matters driven by marketing strategies intended to enhance the portfolios of the stockholders of a defense contractor rather than the long-term security of national and international society.

Clearly, such maneuvers raise serious ethical questions. What is the propriety of defense contractors having direct access, for marketing and promotional purposes, to high-level NATO ministers?

By what right do company officials pin access to the secret deliberations of committees such as the HLG? What is the role of the US government in establishing and promoting such arrangements? These and related questions are of the utmost gravity and relate to the de facto surrender of strategic and foreign policy-making to unnamed business persons well beyond the influence or scrutiny of the US people.

"Decapitation" Scenarios

On October 13, 1981, the opening session of the first of an annual series of unclassified National Security Issues Symposia, designed to present and explain the Reagan Administration's national security policy to a responsible cross section of the lay community, was held in Bedford, Massachusetts. That day, Dr. Richard Pipes, a staff member of the National Security Council, presented a candid assessment of the Administration's view of the Soviet Union. He stressed, in particular, those factors in the development of the Administration's strategic nuclear policy that in his judgment justifiably distinguished it from all previous policies, to wit, that it was the fast to comprehend fully the deterrent value of "counterforce" strategy. Prior administrations, Pipes maintained, had mistakenly judged the Soviets to value life and property as we do and therefore had naively implemented predominantly "counter value" strategies threatening life and property. The Soviets, he contended, value not fife and property but, rather, their weapons, their communications and control systems, and their political system. Therefore and this was the major breakthrough of the Reagan Administration according to Pipes deterrence was best served by so-called decapitation scenarios, in which the Soviet command and control structure is the preferred target.

Any high school student can recognize immediately not only the cultural hubris, but, as well, the fatal flaw in the logic of these remarks: You can't talk to someone you've just decapitated. Nations and their leaders do not exist, like Schroedinger's cat, in some kind of quantum mechanical "superposition of all possible states" that would enable one to communicate with the dead. Later the same day, Lieutenant General Brent Scowcroft, then an Administration consultant on foreign policy and national security affairs, wrestled with the implications of decapitation scenarios:

There's a real dilemma here that we haven't sorted out. The kinds of controlled nuclear options to which we're moving presume communication with the Soviet Union, and yet from a military point of view one of the most efficient kinds of attack is against leadership and command and control systems. Much easier than trying to take out each and every bit of his offensive forces. This is a dilemma that I think we still have no completely came to grips with.

The dilemma fairly screams out that something is dreadfully wrong with our logic. Yet the power of ideology over rational thought is so great that evidently few seem to have spotted the fly in the ointment. As viscerally appealing as their advocates may find them, decapitation scenarios make no sense whatsoever. Should a decapitation ever be carried out, the most likely response would be an uncoordinated paroxysmal release of virtually every strategic weapon in the opposition's arsenal, extinguishing essentially all life on earth.

This indictment of the fundamental illogic that lies at the core of US strategic policy is reinforced by what I infer from an exchange between General Scowcroft and Dr. Edward Teller (former director of the Lawrence Livermore laboratory and considered by some to be the father of the hydrogen bomb) at the National Security Issues Symposium in Bedford, Massachusetts, in 1981. Dr. Teller asked two questions of Scowcroft, the second of which was as follows:

There is a second question which I'm afraid we cannot, for many reasons, discuss in detail. But I would at least like to mention it. Has it been considered to concentrate the power of the President and his various successors [in the event of decapitation by the Soviets] on various levels not only on commanding an attack a counterattack -but on restraining the usefulness of the weapons in such a manner that this restraining power automatically lapses unless renewed periodically? And it also lapses automatically on the unmistakable signs of actual explosives hitting the United States? What I have in mind is a hierarchy where it does not pay the enemy to attack the top leadership because all they do thereby is lose the peoples with whom they can negotiate; and instead they have an interest in preserving our leadership. Has that been considered?

Replied General Scowcroft: "You are very correct Dr. Teller, we can't talk about it. Is that enough of an answer?" To which Teller responded: "A little more than enough. Perhaps I have to apologize about mentioning that such an idea could even exist."

My reading of this exchange is that the United States very likely has in place some automatic retaliatory response to be activated in the event of a confirmed attack on US soil or, most certainly, a confirmed loss of the US National Command Authorities. If so, the astonishing conclusion is that, if the much trumpeted deterrent value of our force structures and strategies fails, the United States will with malice aforethought, commit national suicide along with the massive collateral humanicide of most, if not all, of the rest of the world's population. This is, I believe, no less than an intent to commit a crime against humanity of the highest order and it therefore merits the strongest possible condemnation. That any nation even entertains such an idea should be a matter subject to the jurisdiction of the International Court of Justice. Yet, to my best knowledge, this exchange, which took place in the presence of reporters from at least two major US newspapers and later was published in the official transcript of the National Security Issues 1981 Symposium, never was reported to the US people, never was brought up before Congress, and certainly never became a subject of national debate. Besides being a remarkable example of the selective inattention of individuals of all persuasion, however, it demonstrates again the extreme misapplication of the technologically feasible in the name of national security.

At a minimum, the exchange lends credence to the proposition that we are moving closer and closer to automated responses that may function after our demise and that do not serve the true national interests of the living. As such, it raises a host of troublesome questions. What if the attack on the United States comes from a country other than the Soviet Union? Bearing in mind that it takes weeks even months to sort out the precise causal sequence of events in disasters far less significant than a nuclear war, might we unleash the full fury of our nuclear arsenals upon the Soviet Union in response to an attack from a Third World country? How discriminating is an automated response under fire likely to be? And what is the likely Soviet response to the knowledge or suspicion that we have such a plan? When pressed on the logic of targeting Soviet political and military command and control centers, many, including General Scowcroft, have stated their belief or hope that the Soviets would not respond paroxysmally if "decapitated." And perhaps, today, in the glow of glasnost and perestroika, there is some justification for this belief or hope. But if the United States, the alleged "good guys," have already contemplated such a plan, who can be sure that the Soviets, not long ago branded the "focus of evil" in the world, will not do likewise? The quality of thinking here is manifestly deplorable!

The Reagan Administration's analysis, as articulated by Dr. Pipes, is a textbook example of the technique of deprecating the opposition to justify the undertaking of otherwise unwarranted decisions and actions. In this case, the allegation, accepted carte blanche, that the Soviets "eat babies" was used to justify and motivate the massive arms buildup initiated in the first Reagan term. No one questioned seriously whether Dr. Pipes really knew what he was talking about; whether he or others might not have had a hidden agenda; whether the claims about the Soviets would stand the test of critical examination; and, in any case, whether the Administration's proposed responses were in any way appropriate or beneficial. The less reasoned the exhortation, it seems, the less reasoned the response.

Application of Technology to Security

Hiroshima-Napsaki, technology has been misapplied in the name of national and international security. Space permitting, each could be supported by many more illustrations and analyzed in far greater detail and the areas of principal concern themselves could be extended considerably beyond those I have chosen. However, the material I have presented is already sufficient to identify unambiguously the broadest and most fundamental reforms that are required if we ever are to find lasting and stable security without relying on nuclear deterrence. Our hope cannot continue to reside with an indefinite series of technological fixes that become exponentially ever more complex and bizarre.

The first necessity and, unfortunately, the one that will be far and away the most difficult to achieve, is a revolution in the way people think of themselves and their country; in their general level of interest and knowledge of domestic and foreign policy; in their tolerance of alternate viewpoints both from within and without; and, above all, in their willingness and commitment to be faithful to notions of human solidarity. Americans, for example, should take the foreigners' image of the "ugly American" seriously and examine it for its true meaning and content. Everyone can learn much by listening carefully to their detractors. In the final analysis, genuine security will be derived from the genuine respect not fear of others.

Of course, human nature being what it is, this sort of revolution cannot be legislated into existence. We are talking about the way people think, and that is ordinarily very much dependent on the way we think those around us think. Herein lies the power of jingoism. People are exhorted to think uniformly and to derive their sense of the correctness of their thoughts from the reinforcement of those around them, rather than from self-examination and reflection. The latter traits are not seen as virtues and those exhibiting them are judged to be out of the mainstream, a place where, presumably, no one wants to be. Therefore, if societal thinking is to change in the fundamental way I believe it must, it probably will require an enlightened charismatic national leader. Though he still must stand the test of time, Mikhail Gorbachev may well prove to be such a leader.

Of the areas that are amenable to legislative action, the most wanting is the need to institutionalize, with an eye to clearly defined objectives, the process whereby technology-security decisions are made. First, governments must deliberately insulate the making of strategic policy from the vagaries of political fashion. Second, they must assure that a comprehensive cross section of thought -political, legal, moral, technical is brought to bear without prejudice upon all policy decisions. The process must be both flexible, in the sense of admitting the greatest possible range of analysis, and rigid, in the sense of being essentially immune to political caprice. Policy must square with the highest moral principles, withstand the most rigorous legal analysis, and be feasible in the technical sense, both short- and long-term, with a full accounting for all conceivable contingencies both favorable and unfavorable. Strong and effective measures must be instituted to minimize to the greatest degree possible the influence of parochialism and the squalid divisiveness of such base tactics as the dehumanization of other peoples.

The standard called for transcends any that can reasonably be expected of an elected body or official. I have in mind the establishment of national strategic policy-making bodies that are as free from political influence as, say, the judicial branch of the US government, but which, unlike the US judiciary, play an active rather than a passive role. Policy, and its realization, must be stable over the long term and must be consistently the product of the best thinking that can be brought to bear. There must be virtually no possibility that debacles such as MIRV and its chain of foreseeable consequences can come to pass. Likewise, exhortations that dehumanize the enemy and intellectual nonsense such as the decapitation conundrum never must see the light of day. In the United States and elsewhere, the executive and legislative branches of government, including the defense establishments and relevant others, must be obliged to carry out policy established by such bodies with failure to do so being regarded as a transgression of the highest order. Through such bodies, the quality and consistency of strategic policies, as well as their implementation, can be elevated to levels commensurate with the complexity and danger of the problems the world is likely to face in the foreseeable future.

Obviously many lower level and, hence, more specific, reforms are called for as well. The first of these should probably be a sweeping reform of the way we solicit, develop, and procure weapons technology. As suggested by the example of the Pershing 11 missile, there must be strict laws limiting access of individuals and corporations with vested interests to policy-makers and their deliberations. Penalties for violations should be severe.

We need also to treat weapons-making for what it is: the lamentable diversion of human talent into the creation of sophisticated means to kill and mutilate fellow human beings. However this pastime may be justified, its moral implications must never be allowed to drift far from view. Weapons-making, if it must be, should be seen as a public duty, not as a glamorous high profit enterprise to which business-as-usual ethics and practices apply. To this end, defense contracting should, in effect, be nationalized. Weapons should be made only because it is deemed that they must be made, not because a company or corporation has a stake in a follow-up contract. To the greatest degree possible, the incentives to keep new weapons systems coming for their own and a company's -sake should be eliminated.

Defense employment should be deglamorized for individuals as well. Salaries should be set by law somewhat below comparable positions in civilian industry. This would provide immediate relief to the civilian sector in terms of competitive access to the best engineering and scientific talent (which, in the case of the United States, might also mitigate somewhat the decline of US technology in world markets). It also would have a desirable sobering effect upon persons considering careers in the defense industry. It is to be assumed that as a result of the absence of nuclear weapons and the presence of policy-making reforms there would be a decrease in demand for weapons technology. If, in spite of this, diminished individual incentives lead to a shortfall of available workers, defense employment could be made a form of obligatory national service. After all, if defense employment really is for the national good, and not merely for high profit, high salary, and high glory, then the patriotic motive should suffice.

Predictably, proposals to take such a heavy hand to the defense industry and defense employment will be met with resistance because such actions are assumed to create economic dislocations that can throw tens or hundreds of thousands of people out of work. But this response often morally reinforced by calls to sacrifice for the greater good, notwithstanding that the lives spared by an averted war tomorrow surely are worth the loss of jobs today overlooks the fact that the claim of economic dislocation is without merit in the first place. As Professor Lloyd J. Dumas has pointed out with eloquence, defense employment is essentially a form of welfare. The products of defense employment add almost nothing of value to the economy, where value is measured not in terms of dollars that have changed hands but in terms of goods and services that lead directly to the creation of additional goods, services, and employment. To use a familiar analogy, dollars spent on defense are like paying to have a hole dug in the ground and then paying to have it filled in again; money changes hands, but no real value in terms of the creation of new goods and products or meaningful employment is added to the economy. When measured in this way, there is very little difference between paying a defense worker to do defense work and paying the same worker to stay at home; if one simply continues to pay dislocated defense workers their salaries, as if they were on welfare, society would not be worse off economically than it is right now. On the other hand, if instead, as Seymour Melman has suggested, we were to invest a moderate amount in retraining dislocated defense workers for civilian employment, we could be a good deal better off than we are right now. The talent freed up and properly retrained would constitute an enormous resource that could be applied to the revitalization of our economy and our industrial infrastructure, as well as to reversing the deterioration of health services, education, housing, roads, public transportation, the environment, our inner cities, and other social services and programs all of which have been compromised by defense spending. Such a conversion of resources could lead eventually to enhanced national security, not of the kind guaranteed at the point of a gun, but that serves as an example for the world to emulate.

Finally, I recommend a relatively minor reform that happens to be near and dear to me as a taxpayer and citizen concerned with the state of higher education in the United States. It is now the law in the United States that, to qualify for US government educational loans, male students must register for the draft. This law, which incidentally illuminates well our current priorities, should be repealed forthwith. In its place should be substituted a regulation requiring colleges and universities whose faculties receive technological and related research grants from agencies of the Department of Defense (whether or not the projects funded are classified) to provide mandatory standardized courses on the history and dynamics of the arms race to all degree candidates in fields of interest to the defense industry. It is unconscionable that young people whose talents are turned to the perfection of weapons and weapons-related systems should be largely ignorant of the political and foreign policy implications of their work. No world, whether free of nuclear weapons or not, can be genuinely secure when the generation that inevitably will someday take over the reins of power is ignorant in these respects.

Summary

The terms glasnost and perestroika and the unprecedented changes taking place in Eastern Europe in their name afford a literally undreamed of opportunity to apply the principles of alternative security to the now conceivable prospect of a world essentially free of nuclear weapons. In the spirit of the ancient Chinese curse, "May you live in interesting times," the blessing could devolve into tragedy if we fail to set aside our former ways and embrace the opportunity from all sides. This could well be our last chance to say no to the gut impulse to seek security through technological means. As I argued at the outset, we have advanced the scope and sophistication of technological wizardry to the point of vanishing returns, a condition from which there is no reprieve for reasons that are fundamental and beyond the ability of the human will to mitigate. That this opportunity should have presented itself at this propitious moment is worthy of the accolade "miracle." It would be prudent to conduct our affairs accordingly. Fortune is not likely to look so favorably upon us more than once.

2

Foundations of National Security

Introduction

A wartime economy, including those of eighteenth- and nineteenth-century states, whose armed challenge and response were less ambiguous than they are today, can be viewed realistically as a social mechanism designed to solve the same fundamental problems as its peacetime counterpart. Modern international relationships, containing continuum of peace-war mixtures from active cooperation through protective occupation, satellitization, cold war, police action, and a spectrum of "limited" war, to thermonuclear holocaust, make the application of Clausewitz 's dictum to the economic sector seem almost a triviality.

The basic economic problem that exists in any society, at any time, under any constellation of circumstance, springs from the scope of human desire and the niggardliness of nature relative to it. Imagine the near-infinite number of goals a large, modern nation state could seek to attain had it limitless resources, including all of those goals which directly benefited the whole nation as a collective body, and, in addition, the whole set of his material desires for every individual in the nation.

Either of these two groups of ends the collective or the individualistic would be far more expensive in economic resources than those available to even the wealthiest nation. A resource constraint looms up, enforcing upon society the need to pick and choose its ends. The choice of goals to be achieved with limited means is the essence of the economic problem. It implies that, in some manner, a society must array all of the goals it might seek to attain in terms of their relative desirability. From this ordering of relevant objectives, together with a knowledge of the prospective resource cost of each goal and the total number of resources available, decisions must be reached. How many of its resources should society devote to collective goals and how many should it free to satisfy its members' individualistic goals? Given the resources available for collective goals, how many should be used for building roads and how many for B-58's? What resource effort should be made for the development of "gray area" economies as opposed to building domestic research facilities? Which economies and what types of research should be favored? From the material wherewithal available to fulfill individuals' goals, what goods should be produced and in what quantities? Which individuals should get them? How should these goods be produced: with large amounts of capital resources and small amounts of labor, or vice versa? Where should they be produced: New York, Peoria, abroad?

Ideally, then, society relates the gratification of all relevant objectives collective and every individual's to its own well-being, and obtains the anticipated resource price of each. The need for choice is clear, but what should be the criterion by which this goal is selected over that? A society perturbed by the vastness of the decision-making involved might toss the objectives into an urn, and draw, now this collective goal, now that individual goal, in a random fashion, until resources were exhausted. But chance is the absence of a criterion: used thusly it is social aimlessness and would contradict any of our ordinary notions of rational behavior. Surely the society (again, in the ideal) should seek to maximize its conception of its own well-being: for any two possible objectives costing the same number of resources, it should select that yielding the higher social benefit, and for any two goals yielding the same benefit, that which costs less should be selected.

To perform this huge volume of decision-making all of which is implied by the fundamental economic problem social mechanisms must be built. Some manner of actually arriving at these decisions (in which the members of the society can acquiesce), using the desired criterion, and of effecting them, in both the collective and individualistic spheres of action, must be devised.

Many of the distinctive characteristics of a society are revealed in this process of decision-making, taken as a whole: the goals it envisages as relevant will reflect every tenet in its ideology moral, political, legal, and the rest. Its ordering of these goals' social values, especially those of the collective versus the individualistic, goes to the heart of its culture, and the degree of agreement among its members in the ranking will reflect greatly upon its ability to function. The number of resources available to it springs in part from its natural endowment, but also from its past willingness and ability to develop that inheritance; and the mechanisms for choice reflect the structuring of power as well as the whole governmental framework and the individual's position in it.

In the United States there is an apparent duality in this process of economic decision-making: on the one hand, collective objectives are defined and effected by governmental mechanisms, and on the other, individualistic goals are products of the "market mechanism." This latter is, in fact, a large number of intricately related markets where individuals sell the resources they control, where businessmen buy the resources to produce goods, where these goods are sold to other businessmen and to consumers, and where paper claims upon future goods are sold for present goods. In this vast arena, featuring interaction among buyers and sellers, prices are determined: the prices of resources, which in turn go far in the determination of a person's income; the prices of goods, which help to dictate how much of what products businessmen and consumers desire; the prices of securities, which help consumers decide how much to save and businessmen how much to invest. From this complicated, interdependent system of markets, in which literally hundreds of millions of decisions involving individualistic objectives are made and coordinated, emerges an allocation of the resources available for the individual sector among the producers of the goods, a set of produced goods, and a distribution of these goods among the individuals in the society.

It is one of the most important characteristics of the United States economy for our purposes that by far the greater number of economic resources (about 80 per cent by value normally) are regulated by the market mechanism rather than by government dictate. Ideologically and historically, the market mechanism is and has been the fundamental organizing principle of economic activities in this nation.

The adoption of the market mechanism by a society to control the use of the greater part of its resources has far greater implications than those contained in the purely mechanical aspects of its performance. For the market mechanism carries within itself a set of social rankings or priorities. Most fundamental of all is this dictate: the major end of economic activity is to satisfy the selfish desires of individual consumers, and, therefore, consumers should be allowed to decide ultimately what is produced, how much is produced, what its uses should be, who shall obtain it, and what provisions for tomorrow will be made.

The market system is, then, in more than the mere formal sense, an individualistic system. Inherent in it is the overriding importance of individualistic economic goals and, correspondingly, the minimal importance of collective goals. To the extent that collective goals exist, provision for them must be made in an extra-market manner, since the market is not equipped for such decision-making. The governmental apparatus concerned with these collective needs tends to be looked upon antagonistically, and the implementation of its choices tends to appear an "unnatural" overruling of the dominant economic priorities. Taxation the most important method of asserting the greater social benefit of some collective goals over some individualistic objectives is of necessity resented, since it strikes directly at this most fundamental priority of the market economy. Under such a regime individuals discount heavily the benefits they receive as a group when contrasted to those pleasures they might have enjoyed directly and individually.

This dichotomy of objectives allows us to group most ordinary economic problems of the United States into two types; 1.) quarrels with the priorities scheme resulting from the operation of the free market mechanism within its proper sphere; and 2.) quarrels with the priorities placed upon the individualistic sector's objectives generally as against the collective sector's goals.

In the classic idealized version of the market mechanism power is atomized so that permanent blocs of economic power cannot exist, and ad hoc groups cannot exert an influence out of proportion to the number of individuals in them. The classic liberal function of government, therefore, re the market mechanism, was the laissezfaire task of protecting this atomization of power, or where, in the nature of things, one group of individuals (e.g., children) lacked parity of power with another, to redress the balance with the power of the state. In the realistic market system, however, where permanent power blocs do exist, the need of government to strike a viable balance of power among them to keep the market mechanism on-going becomes most important. Examples of this first type of economic conflict are the refusal of the agricultural bloc to accept the priorities of the market mechanism; the struggles of labor and management over submission to the market's dictates in the distribution of product; the demand of entrenched industry and labor for interference with product flows which the market would bring in from abroad.

The second type of problem, as might be expected from the discussion, is much more challenging to the society's adaptive potential. During periods in which much agitation exists to lift the collective goals' priorities at the inevitable expense of the individualistic goals' fulfillment, much heated debate is likely to be generated. While the first type of problem finds all parties in substantial agreement as to the acceptability of the market framework and the broader lines of its solutions, and while one bloc's power serves to countervail that of the other, it is often true that neither condition holds when problems of the second type are posed. Business, labor, and agriculture may resent any restriction upon individuals' rights to use resources for their own ends, since, given the basic priority inherent in the market, such restrictions are always "revolutionary" in a sense that shifts in power balance seldom are.

Soviet Russia, with the largest "directed" economy in the world, does not wholly escape problems of the second type, but they are viewed much more as we view our first type, i.e., as disturbances taking place in a system structured to tolerate them. To date, the basic Communist quarrel with capitalism's free market is not with the distribution of product which its priorities dictate, as Marxist ideology would emphasize, but rather disagreement occurs because the highest priorities are not given to goals which are collective in nature and therefore incapable of achievement by the market: economic growth beyond what the selfish actions of individuals in a market economy would permit, and the economic implementation of a program of revolution. The subordination of individual economic goals to collective ends of this type has dictated the total subordination of market economy to governmental planning.

Nevertheless, both the Soviet and the American societies face the same problem of choice, with all that we have seen it implies: the definition of objectives, the derivation of a system of priorities to order them, the recognition of resource constraints, the selection of a criterion of choice, and the design of mechanisms to effect decisions.

Nor does the need for both nations to shape a foreign policy and provide for its implementation even in conditions of "hot war" alter the nature of the economic problem. Foreign policy objectives are a subset of the society's broader objectives, which must be considered as having a claim on resources to an extent determined by their relative social importance. The preparation for and waging of war involve the same need for choice by society as the more general case.

Why, then, do we single out a body of problems in the economic arena as problems of national security? Many factors enter into the answer. The sheer magnitude of expenditure now necessary to provide a military posture is certainly one. The hypothetical nature of the future events which justify the need for such posture, or of some quantity or quality of it, and the consequent volatility of such expectations, are others. At the risk of oversimplification, these reasons interacting with another, derived from our discussion above, must be stressed: the great shift of priorities toward the collective goals and against the individualistic ends implied by the provisioning of national security in a market economy.

Modern warfare and the maintenance of peacetime military posture have become the two most important instances of the "revolutionary" impact of collective goals upon a resistant market system. As an index of the magnitude of this encroachment in prenuclear total war, the Federal Government's expenditure on national security rose from 2.2 per cent of the United States' output of goods in 1940 to 41.5 per cent in 1944. These expenditures more than doubled their relative share of national production between 1950 and 1952, during the Korean struggle, rising from 6.4 per cent in the former year to 14.1 per cent in the latter year. The maintenance of our peacetime security for an average year in the period 1930 to 1939 absorbed about one per cent of national product; this share averaged 11 per cent from 1955 through 1957.

We have seen that such diversions from the individualistic sector must be obtained by extra-market decision and by measures which overrule market priorities. The magnitude of them and (during peacetime) their doubtful need, ensure that heroic measures will be required to effect them. Most of the economic problems implied by national security hinge upon this "diversion": ways of reversing the society's usual and inherent bias toward individualistic economic goals; the size of the needed diversion; the most efficient methods of effecting the diversion; the advantages and disadvantages of alternative methods of effecting the diversion; the phasing through time of the need for and production of the diverted goods; the mechanics of procurement of the needed goods; and so forth.

We will therefore concentrate upon this diversion process. The analysis will proceed by studying these problems at three levels of intensity: total war, limited war, and peacetime preparedness. For the economist's purposes, the term "total commitment" is well-suited to describe a conflict to whose objectives a good-sized portion of a nation's stock of resource wealth is devoted. The word "commitment" has two meanings in this context: 1.) the temporary use of the services of a stock of resources, and 2.) the subjection of such wealth to the possibility of destruction as well. For reasons which will become apparent, our use of the phrase "total commitment" in later sections will describe struggles in which a substantial portion of the economic resources of nations is subjected to destructive attack. But, temporarily, we shall adopt the broader meaning of the term, which includes the mere use of large amounts of resources.

The role of strategic warfare in total wars is likely to be great. It consists of a body of techniques for interfering with an enemy's use of economic resources, either by lessening the effectiveness of their services, as in the denial of raw materials or food by blockade, the reduction of labor efficiency by terror bombing, etc., or by destruction of the resources, as in the case of strategic bombing.

The preconditions for the re-emergence of wars of total commitment in 1914-1918, after a century of limited hostilities, sprang from four major streams of social development: 1.) the rise of the citizen army in the American Revolution and the extensive use of conscription in the Napoleonic period; 2.) the capacity for mass enthusiasm and response to patriotic-nationalistic appeals in the new secularist societies; 3.) the development, in large part as a by-product of the technological revolution in peacetime industry, of new weapons and logistics systems, giving armed forces greater firepower, accuracy, and mobility; and 4.) the emergence of the modern, wealthy industrial nation with its large surplus product over the needs of subsistence. Total commitment depended upon the development of "mass": mass armies, mass enthusiasm, mass laughter technology, mass-production economies. The first two had arrived by the end of the Napoleonic Wars; the emergence of the third we might date conveniently at the time of the American Civil War; but for the fourth that one in which the economist is most interested the world had to await the maturity of Western industrial economies. The tragic date of 1914 marks this coming of age. The significance of this fourth precondition for modern total war is best judged against the American and European experience that seemed germane on the eve of World War I. For example, during the American Revolution peak mobilization occurred in 1776, but the young nation freed only about 3 per cent of her population for service under arms, a percentage which declined to 1.5 per cent by 1780. During the Civil War the peak strength of the Southern army (1863) plus that of the Union armies (1865) totaled only 2.8 per cent of the nation's population. The greater bloodshed and destruction ascribable to the advances made in weapons and communications yielded tragic human losses about 2 per cent of the population became war dead and, of course, created a cumulative strain on resources, but at war's height the nation could free less than 3 per cent of her population.

As World War I approached, Great Britain, then at the height of her empire, was spending only 3 to 4 per cent of her national income to maintain a military posture. On the eve of war, the French general staff estimated that approximately thirty factories employing 50,000 workers could produce the munitions needed for its armies. And estimates for the total cost of World War I, made in its period of imminence, varied from $18 million per day to $55 million per day.

The experience that seemed relevant to American and European statesmen in 1914 led the world to World War I with no understanding of the tragic sustaining power of modern industrial economies. In this sense, World War I was a war of unforeseen and unprepared-for total commitment. From its diversion of 3 or 4 per cent of production in 1913 for military posture, Britain increased this proportion to 46 per cent of an even larger product, in the period of its greatest effort an absolute rise of 1150 per cent in real resources. The modern economy had reached such levels of production that almost 50 per cent of its resources could be devoted for long periods to collective goals of this type. About 6 per cent of Europe's population suffered death from war causes in the conflict, and 33 million civilian and military casualties resulted. About 61 million men were mobilized at one time or another at costs estimated at $338 billion of resources in 1919 dollars, or about $163 million per day. Included were those for a French effort employing 1.6 million workers in munitions in 1917, instead of the 50,000 envisaged in 1914.

During the following two decades, statesmen might underestimate the extent of the commitment required for a struggle between major blocs only at the risk of their own nation's political survival. But, given their objectives in foreign policy, and the implications of these for military posture, weapons technology offered them an option. They could phase their armed strength through time, adopting a shallow or a deep time structure for their nations' military postures. For example, suppose a nation decided to devote $1 billion of its resources to increasing its military strength. It might increase the present size of its armed forces and thereby enhance its power-in-being. This type of power has a shallow time structure, since it can be fielded in the nation's causes with little time lag. On the other hand, the nation might decide to invest its resources in aircraft and munitions plants, enlarging its military posture, but structuring it more deeply in time, since, from a given point of need, it would take a longer time to make such power effective than it would for the first type. Lastly, it might elect to build steel mills and aluminum plants, which would result in a still deeper time structure for its national power.

It is one of the more important tasks of economic statesmanship to structure the nation's military power through time in such a way that, given society's collective goals in the foreign policy field, they can be achieved with minimal resource expenditure. A nation whose economy is dominated by the free market mechanism will tend to be biased toward the deeper time structures for several reasons. First, such nations, with their stress on individualistic uses for resources, are not likely to be aggressive powers seeking ambitious collective objectives. Second, deeper structures are cheaper, both to build and to maintain. For example, a society with the greater part of its strength dormant in its industrial potential will have very few costs on military account, since the steel mills and durable goods factories will yield potential military posture as a costless by-product of their peacetime activities. Third, and for the same reasons, such a posture minimizes the interference with the fundamental priorities existing in the market mechanism, and thus minimizes the conflict between collective and individual goals. And, lastly, a deeper time structure is more productive. That is, if steel mills and aluminum plants are built instead of aircraft factories, when war comes the former make possible more of the latter -given time. The enhanced productivity is achieved only at the expense of having to wait for it.

During World War II, and the preparation for it, it was possible indeed, vital to build temporal depth into nations' military postures. It seems certain that Hitler's inferior economic implementation of his ambitious political objectives will become one of the classic cases of failure to structure too few resources properly. The struggle between Hitler and General Thomas, chief of economic planning for the high command, on the specific question of time structure, with the latter insisting that war before 1945 or 1950 would find Germany's strength insufficiently deep, was resolved in favor of Hitler's contention that a shallow structure of restricted dimensions would prove equal to the tasks. In many respects, even after the final refutation of Hitler's doctrine at Stalingrad, such depth of posture that sprang from the unprepared German economy was never properly mobilized for the war effort. More realistic Japanese statesmanship frankly recognized that, with an economy whose output of products was roughly 11 per cent of United States product, Japan's only chance of victory lay in achieving large forces-in-being relative to her foes (by dint of great economic resource diversion and surprise attack upon Allied naval power) to be used for fast advance into the South Pacific and to erect a naval defensive wall, with the hope for a negotiated peace. The Coral Sea and Guadalcanal engagements were the Stalingrads of Japan's leadership the sign that containment of her shallow military strength had been achieved by her enemies, whose deeper structures now presented the prospect of sustaining a long war Japan could not win.

In 1939, drawing upon the experience of 1914-1918, a British economist wrote: "War, nowadays, is an industrial proposition. It is more influenced by the science of economics than by the art of strategy. The present war will not be won on any playing-fields, at Eton or elsewhere, but in the mines and workshops of a thousand grimy industrial towns.

Perhaps this economic aspect of war is not quite so novel a feature of our own times as we like to imagine Britain's wealth has always helped her to win her wars. But it has certainly never been recognized before the present. It was the chief military lesson of the war of 1914-1918."

The lesson was repeated in World War II: economic depth overcame the initial challenges of shallow military power and ultimately overwhelmed it. But what of tomorrow? Is it possible to project this lesson into future wars of total commitment as confidently as was done for World War II? Does "economic potential for war" still have the meaning it had two decades ago? These questions challenge both the usefulness in future total wars of the strength derived from deep-structured military posture and the ability of a nation to bring it into play. We shall assume away the first problem, on the realistic grounds that a nation must prepare for wars of limited commitment today, and, therefore, such strength will have an independent desirability. But the second question remains. To answer it, we must revert to the concepts of the "diversion process," "total commitment of resources," and "strategic warfare "We may dispose of the implications of the first of these terms quickly, but we will shortly discuss them in different contexts at greater length. The capacity to exploit economic depth depends importantly upon the ability of a society to reverse the order of priorities given social objectives as opposed to individualistic, and to keep it reversed. In two world wars, the United States has demonstrated that under conditions of total involvement the market mechanism's priorities can be so changed with minimum interferences and controls. Moreover, with these minimal constraints, it can continue to be used as a decision-making mechanism. Appeals to patriotism and self-sacrifice have been sufficient to effect this transformation and to maintain it for long periods, and we should have few doubts on this score for the future. But significant changes have occurred in the meanings of the other two terms. They are best considered under three question marks:

1. Are the conditions of a future total war likely to present any changes in the existence of economic targets which will make them less attractive to strategic warfare techniques? Modern economies are concentrated economies. For example, the area of the United States bounded by the Atlantic Ocean, the Mississippi, the St. Lawrence-Great Lakes System, and the line of the Ohio Potomac Rivers, contains about 50 per cent of the nation's population, 70 per cent of its manufacturing employees, and produces about 70 per cent of the value of manufactured goods for final use. In 1952 it produced 77 per cent of our primary metals, 77 per cent of fabricated metal products, 83 per cent of our non electrical equipment, 87 per cent of the output of electrical machinery, 70 per cent of the transportation equipment, and 75 per cent of our rubber products. Similarly, in the Soviet Union, the Fertile Triangle, whose corners are drawn at Leningrad, Odessa, and Irkutsk on Lake Baikal in western Siberia, defines an area about equal in size to that of the United States east of the Mississippi, within which occurs 90 per cent of her population and farms. Almost all of her economically useful land, and, more roughly, all five of her major industrial areas Leningrad, Moscow-Gorky-Yaroslavl, the Ukraine, the Urals, and Western Siberia are contained within it.

It is needless to say that economic concentration to this extent reflects strongly the attempt of societies to minimize the resource costs of their production: steel is not produced in plants with 1,000-ton capacities, nor popular automobiles in firms whose outputs are 10,000 units per year, because the economies of large-scale production do not appear at these "dispersed" levels. Similarly, firms tend to agglomerate in spatial clusters in order to draw upon common pools of resources, or specialized skills, or for the convenience of customers. Although it is understandable that governments should seek to encourage new plant locations in cities of less than 50,000 population separated by ten miles of open country, the failure of such programs was predictable in the absence of substantial and continuing subsidies to compensate firms for the costs attendant upon ignoring the market mechanism's priorities for the sake of collective objectives.

Moreover, even were the dispersal of production facilities to be successful, another consideration would nullify its strategic meaning. Transportation systems are by their nature not dispersible, for they are channels of movement between nodal points of concentrated activity. These channels and nodal points are peculiarly vulnerable to strategic bombing: even under the severely limited capabilities of such bombing in World War II, the German transport of goods was seriously restricted by a seven-month attack on rail facilities. In the spring of 1944, a two-and-a-half-month attack against ninety rail yards in France reduced traffic to 13 per cent of its prior flow, and, in Italy, fighters and medium bombers alone were successful in interdicting almost all rail traffic with only 12,000 tons of high explosive. Nor must it be thought that the vulnerability arose from the existence of possibly dispersible freight yards: on the contrary, the most effective attack was upon the line rather than the yards. On the basis of such experience, the Strategic Bombing Survey in Japan concluded: "It would have been possible by concentration of effort from April (1945) on, to drop, say, 10,000 tons of (high explosive) a month on railroad targets. From experience in Europe and the relative size and vulnerability of the Japanese system, it might have been expected that under a bombardment of this magnitude, traffic would have been reduced to disastrous levels within about two months."

Therefore, any significant dispersion of production facilities is likely to place severe burdens on secondary transport networks in times of limited war and increase the resource cost of peacetime production. It seems likely, as a consequence, that modern industrial powers will continue to be highly concentrated spatially.

What are the implications of changes in the ease of access by belligerents to such economic targets? The emergence of longrange bombers, the development of refueling techniques, and the promise of intermediate and long-range ballistic missiles launched from land, sea, and air, coupled with the lagging development of active defense, lead to the conclusion that access capabilities will probably continue to improve.

Given the continued existence of economic targets and improved access to them, what will happen to the capacity to destroy them? The revolutionary breakthrough in destruction technology seems to relegate to the limbo of irrelevancy the concept of war potential in depth for wars of total commitment. The destructive power of modern warfare with nuclear weapons yields the unquestioned technological ability to destroy these targets at the outset of hostilities.

On the basis of the unclassified data available, the following is a set of conclusions about the economic effects of a surface burst from a 20-megaton nuclear weapon on a typical large American city. The attempt was to obtain firm minimum estimates of such impacts, not realistic potentialities of the weapons. Therefore, the following conclusions are most conservative, though obtained from the most recent damage analysis.

The average area of the ten largest American cities (excluding New York, which presents problems sui generis) is about 100 square miles, and contains a population of 1.4 million. Were this area circular, it would have a radius of about 5.6 miles. If a 20-megaton weapon were dropped on the center of the city, the following damage could be anticipated:

1. All blast-resistant, windowless, reinforced-concrete buildings within a circle of 4.5-mile radius would be rendered useless;
2. All multistoried, massive-walled buildings would be rendered useless within a circle of 7-mile radius;
3. All multistoried buildings with reinforced-concrete frames and small window area would be useless within a circle of 8.5-mile radius;
4. All brick apartment buildings would be useless within a circle of 11-mile radius;
5. All one-storied industrial buildings with heavy steel frames would be useless within a circle of 12-mile radius; with medium steel frames, within a circle of 15-mile radius; with light steel frames, within a circle of 18-mile, radius;
6. All telephone and power lines would be completely destroyed within a circle of 12- to 15-mile radius.

At the most optimistic, 136,000 deaths from direct and indirect blast and drag forces and from thermal and initial radiation effects would occur, with perhaps another 140,000 badly injured. Fall-out casualties are difficult to estimate because of the many variables involved, but let us arbitrarily increase the casualties to 300,000 to account for its contribution to death or disability within the city.

With 20 per cent of its population dead or disabled, many others too ill physically from radiation sickness to participate in any effort, most of the remainder forced to take cover for 24 to 72 hours, its survivors homeless and with no safe food supplies, help from its hinterland reduced by the fall-out danger, and all of its collective capital destroyed, the city could no longer be considered an operational entity. There would simply be no mechanism to function. This helplessness is compounded as the number of cities so attacked grows, and national and state coordination potential also disappears. "Broken-backed war" seems a remote possibility indeed.

For the first time in modern history, strategic warfare has become large-scale, non attritional warfare. Industrial facilities cannot be counted upon to provide the means of their own repair; they can be destroyed at one blow. The lesson is plain: total commitment of an economy's resources today goes beyond its World War II meaning to become the commitment to total destruction. Depth in military posture is meaningless in this context, so that time structure must be shifted forward to the limit for potential to fight this type of war. Force-in-being is actively employable as deterrent force, potentially useful as spiteful retaliation. Force-to-be, measured in some calculus of steel mills and wheat-fields, must be given a zero weight for this type of warfare, although it must be given an important weight in building recuperative potential from nuclear attack.

By virtue of the destructive capability of the new technology, total commitment, in the sense of catastrophic destruction of a nation's resources, becomes a new absolute, and wars which in former times would have been termed total would now fall in the less-than-total category. The touchstone by which we shall distinguish wars of limited commitment is the prospect of a nation's economy sustaining the weight of destructive attack without having its resource base rendered useless or nonexistent. In this sense, World War II was a war of limited commitment, despite the totality of its objectives and resource use.

One of the advantages of this classification is that it militates against the identification of "limited war" with the Korean War. Future wars which do not pose a threat to the existence of our internal economy may take on much larger dimensions than the latter. The relatively short duration of the Korean War, its maximum demand of only 14 per cent of our resources, its occurrence immediately after a period of consumer "stocking-up" on durable goods, and the ability to derive the resources needed for its prosecution in largest part from the normal growth increment of our economy, made of it the first completely tax-financed war of our history an important index of the economic mildness of a war in a market economy. But it revealed disturbing portents of future wars of limited commitment: the rates of consumption of equipment by the armed forces were much higher than in World War II,. and the drains of stalemate, even without attrition, were demanding. Could not another war on the Eurasian periphery easily demand resource diversions of 30 or 35 per cent? The most important economic challenge to our society of such wars lies in the need to overrule fundamental market priorities without the support of emotional appeals possible when the nation's existence is threatened.

These wars are likely to result from the wedding of policy and force in such ways that appeals to patriotism and self-sacrifice are only minimally possible. In such an atmosphere, the ability of statesmen to obtain sufficient resources for the pursuit of collective objectives in whose high order of priority they believe may be restricted to the point where the design and implementation of national strategy are seriously endangered. We have seen that the priorities of a market system lie deep and are strong, so that a substantial degree of acquiescence in collective goals must exist before the minimal controls we spoke of earlier can divert the needed resources. In particular, the workability of such controls systems depends upon the willingness of major economic power blocs to find a balance of power and to acquiesce in the degree of sacrifice necessary to maintain it. Certain aspects of the Korean conflict are not encouraging in these respects. A hesitant government delayed instituting a general price "freeze" until the end of January, 1951, after the two major forward-buying waves by consumers and firms in July-August, 1950, and January, 1951, had sent retail prices up faster than they rose during any comparable period in our history except for the six months following V-J Day. Wage stabilization was tied to the cost of living, so that the full wage-price spiral was completed. The conflict between the United Steelworkers and the steel industry, which led to a seven-week strike, an unconstitutional seizure of the industry by the President, and the resignation of the Director of Defense Mobilization over the industry's demands for a price rise in the face of suggested wage increases, which embittered the relations of unions and management throughout the economy, and which occurred in spite of the insistence of the Secretary of Defense that such action would interfere seriously with the prosecution of the war, were symptomatic of this malaise. If such lack of support and hesitant economic leadership can occur in a war when consumption of peacetime goods actually increased on a per capita basis and investment in plant and equipment was roughly constant, more serious wars may raise difficult diversion problems. What are the methods by which a government can effect this needed diversion in periods of limited war? The basic controls needed to override the market's set of priorities are of three types:

1. Those aimed at the reduction or cancellation of the consumer's and the firm's disposal over the economy's resources, either by reducing the amount of their purchasing power, their willingness or ability to spend it, or its effectiveness relative to government purchasing power. The four methods relied upon for accomplishing this task have been taxation; urging voluntary saving upon individuals and firms to free resources; using legislation to force them to save; and, lastly, employing the government's right to create money with which to enter markets and simply bid away the resources from consumers and firms whose money resources are limited.

   This first type of control is designed as a "general" restraint to obtain control of resources: it is meant to affect the consumer's and the firm's control over markets generally, not specifically. But, in limited wars, certain scarce materials are likely to remain in great demand despite any practicable reduction in the means of obtaining individualistic goods. In World War II and the Korean conflict, the most important of them were steel, aluminum, and copper. Therefore, the second component of controls systems comprises:

2. Those for the direct allocation of scarce goods and resources to the achievement of collective goals. We include in this set the rationing of consumer goods, for with certain of them, even if consumers and firms have been reduced drastically in the ability to command goods generally, excess demand will exist and require a formal scheme for allocation to eliminate it if the goods are important to the collective effort.

   Lastly, even with the most extensive controls of types 1 and 2, market forces, particularly in the resource markets, will still attempt to exert their pattern of priorities via price changes. Such price changes, if allowed to occur, can interfere with the allocation of resources required by the collective effort, and can have deleterious effects upon the value of money. Therefore, a third group of controls is needed:

3. Those imposing restrictions on price changes in order to constrain the market's attempt to allocate goods and resources by its own criteria, when such allocation interferes with the collective priorities of the war effort.

The need for and success of the second and third sets of controls depend in large part upon the effectiveness of the first set: the more effective the program to reduce the "voting power" of the seeker after individualistic goals, the less will be the pressure upon scarce goods and upon prices. This first set, therefore, is crucial. The desirable combination of the four methods under this heading must be defined in the light of four criteria: 1.) its effectiveness in reducing the resources used for individualistic goals in order to free them for collective objectives: 2.) its effect upon the incentive of members of the society to produce the needed goods and services; 3.) its legacy of undesirable economic or social effects after the termination of hostilities; and 4.) its distribution of the real burdens of the collective goals among the individuals and groups of the economy.

Taxation, voluntary saving, and the creation of money (by borrowing from the banking system) were the methods used by the

United States in World War II, while only the first two were employed in the Korean conflict. It may surprise the reader to learn that only about 52 per cent of the resources taken by the government in World War II were obtained by taxing individuals and firms, while about 32 per cent came from voluntary purchases of bonds by such units, and 16 per cent were obtained by bidding them away from the market with newly created money. Although we may criticize policy-makers in World War II for this overly modest taxation, no government today would attempt to finance a war of this size from taxes alone, for fear of interfering too severely with production incentives. But even this degree of taxation is a great advance over former wars when the aim was to isolate their impact upon the market economy by borrowing the greater part of the resources or by turning the printing presses. For example, the United States financed only about 20 per cent of the Federal effort in the Civil War by taxes; Germany financed almost 100 per cent of her diversion effort in World War I by borrowing; France, too, obtained less than 1 per cent of the net diversion caused by World War I by increased taxes; while Great Britain and the United States were exemplary in that conflict by imposing taxes of 23 and 33 per cent respectively of the governments' war expenditures.

We cannot enter into intensive analysis of each of these diversion methods, and must limit ourselves to indicating the major advantages and disadvantages of each. It is difficult to contest the conclusion that taxes should be the major reliance of any diversion program in wartime, since they are most effective in cutting into resource use by individuals and firms, they do not lead to huge stocks of liquid assets which can cause inflation at the end of the war, and a good case can be made for their ability to distribute the burden of the war as equitably as possible. On the other hand, their use must stop short of the point where they begin to interfere seriously with incentives. Voluntary saving, given an adequate tax program, is important, but the strength of market priorities is likely to make it incapable of yielding the remainder of the resources needed. Moreover, it does lead to huge liquidity positions for firms and persons, which will disrupt markets and prices after the war. Also, it tends to place a lesser burden upon those whose preference for future goods as opposed to present goods is very strong. Forced saving has not been tried extensively in democracies, but it, too, possesses a disincentive effect which might be almost as severe as that caused by taxes, piles up liquid assets, and places lesser burdens upon those who normally save a large amount or who have large stocks of assets they can liquidate. Lastly, creation of money leads to price increases, increased money supplies, and unequal sharing of the burden of war, but its hidden effects, and, therefore, its minimal interference with incentives, make its judicious use almost indispensable for large diversion efforts. These considerations, and many more we must leave unmentioned, make the construction of a balanced program for the wartime diversion effort a difficult task indeed. But one point must be stressed: it can never be painless. For example, in 1954, 94 per cent of the income-tax payers of the United States paid only 53 per cent of the taxes, but, with those whose incomes were too small to pay any such taxes, they controlled about 75 per cent of the resources that went into consumption. Obviously, in periods of war, when government expenditures must expand at the expense of civilian desires, this pool of resources must be cut into substantially. Regressive taxes, in the sense of those that bear down harder proportionately on the poorer classes, become a necessity in periods of large-scale war. Lastly, a word must be said about the changing character of the diversion process as a war of limited commitment progresses. A war economy passes through four stages;

1. the period of initial call upon the temporal depth of the economy to build the plant and equipment necessary to obtain power in being;
2. the period of initial equipping of the armed forces using the facilities from the first period to produce them;
3. the period of maintenance and replacement of the armed forces' equipment;
4. the period when the end is in sight.

The total size of the diversion effort and the kinds of resources needed will differ in each of these stages, so that there is no "war economy" which can be analyzed independently of its time-phase.

Between 1949 and 1955, Great Britain spent an average of 8.2 per cent of her national product on her military posture, a total resource commitment of perhaps $24 or $25 billion. This latter value is placed in better perspective when it is seen to be roughly equal to that spent by Hitler in resources on the 1934-1939 buildup of his military forces. The expenditure of this sum enabled Hitler to build a war machine that threatened to dominate the Continent and overcome Britain. But when, in July, 1956, Britain was confronted with seizure of the Suez Canal, she literally did not have one division to dispatch to the scene. Nor, four months later, when the ill-fated thrust against the Canal was launched, could she be said to have accumulated overwhelming force to answer what must surely be ranked as a minor challenge.

What has happened to make the maintenance of military posture so much more burdensome today than in the pre-World War II era? Many factors have contributed, of course, but among them is the increased cost of the new weapons and delivery systems, not only for wars of total commitment but for limited wars as well. Also, the shift forward in the time structure of military posture makes an independent contribution to enhancing the burdens of preparation, for instead of stocking "airplanes" in the form of steel mills and aluminum plants, cost lessly, these must now exist before they are meaningful in total war. Moreover, the premium on fast movement of forces has pushed the time structure of limited war forward as well, requiring major powers to carry forces-in-being for "brush fire" operations. And, lastly, the weapons and delivery systems used in both types of conflict, for technological and strategic reasons, must remain independent, with little interchangeability. All of these factors, and others, have combined to increase the dimensions of resource diversion in peacetime to such an extent that near-revolutionary breaks with past thinking are required to accept it.

For, if the diversion problem creates tensions in a market economy during periods of limited war, it is not difficult to imagine how tensions are compounded in peacetime. "Flexible commitment" in the United States since the end of World War II has been, in fact, the struggle of a nation, committed to the priorities of a market economy, to grasp the reality of its new role in the world power complex and to accept its implications concerning the provision for collective goals of national security. The actors might be President Truman putting a $15 billion ceiling on defense expenditure in 1949 or General Bradley testifying that defense expenditures of $14.5 billion in 1949 threatened the collapse of our economy, or Secretary of Defense Johnson embarking on an economy drive for defense spending and touching off the B-39 controversy, or President Eisenhower urging "balanced defense" as the basis for budget cuts to unbalance our forces, or Secretary of the Treasury Humphrey insisting on the aesthetics of a balanced budget, or Secretary of Defense Wilson marking his disinclination to finance basic scientific research by asserting his disinterest in what makes the grass grow green. Underlying the cast is the plot: statesmen struggling against a new order of priorities by reasserting the rectitude of unalterable market priorities.

This soul-searching has resulted in an erratic path for defense expenditures since the close of World War II. Major national security spending, which includes Department of Defense expenditure, foreign aid of a military nature, atomic energy program costs, and the burden of maintaining stockpiles of strategic and critical materials, has gone through one-and-a-half cycles. Falling from a level of $43 billion in the fiscal year 1946 to a depth of $11.8 billion in fiscal 1948, it rose to a peak of $52 billion in fiscal 1953, fell again to $41.8 billion in fiscal 1956, and re attained the $44 billion mark in fiscal 1957 and 1958. Although it reached $46.4 billion in fiscal 1959, it is expected to fall to about $45.6 billion in fiscal 1960 and 1961.

Although these most recent expenditures are quite large in absolute magnitude, and absorb about 10 per cent of our national output, it is difficult to see how they could endanger the national economy. The full employment national output of the United States is currently about $500 billion per year, yielding a per capita product of about $2,900. No nation in history has ever attained such heights of economic performance; e.g., the Soviet Union's present product is perhaps 40 to 45 per cent of American, and its per capita product no more than 33 per cent of ours. Typically, about 65 per cent of our final production is taken by the consumer and another 15 per cent by firms for maintaining and increasing the capital stock of the nation. The remaining 20 per cent is split about evenly between security and non security functions of government. Thus, about 80 per cent of our resources are devoted to individualistic pursuits.

There is danger in any diversion process that interferes a great deal with the 15 per cent investment share, since much of the ability of the economy to grow an ability which, as we shall see, has a peculiar importance for the preparedness economy is dependent upon it. But note the degree of leverage which the sheer size of our national product gives the policy-maker. For every reduction of 1 percentage point in the share of product going to consumption, a total of $4.5 billion in goods is freed for security purposes. The Rockefeller Report's recommendation to increase annual security expenditures by $9 to $12 billion could be achieved with a reduction of 2 or 3 percentage points in the national consumption. It is difficult to see how such marginal adjustments in the relative spheres of collective and individualistic objectives could threaten our rate of growth, or, if increased taxes are used to obtain the extra goods, lead to inflation or to reduction of production incentives.

However, to view the diversion, process against the timeless backdrop implied in this analysis is distorting. The size of our national product does not remain constant, and by virtue of this an alternative method of obtaining increased resources for national security exists. It is to divert product from the normal annual growth increment over a period of years until the desired increase is obtained. This method allows the individualistic sector of the economy to grow along with the collective, lessening the frictions between them. The size of the forward thrust given by growth to economic product should not be underestimated. We have experienced a growth in national product of about 3.5 per cent per year from the third quarter of the last century to the present. Therefore, at present levels, we should expect our national product to grow by about $17.5 billion per year. If national security obtained its proportionate share of this increase, the product available to it would rise by $1.75 billion per year.

To take another case, suppose consumption were allowed to rise by 3 per cent per year, yielding an increase in per capita consumption of about 1.4 per cent per year; investment were allowed to rise at 3.5 per cent per year; and governmental non security expenditures were to rise at 1 per cent per year. This simple model, beginning with the distribution of product between consumers, investors, and government given above, and assuming a national product of $500 billion, would increase annual security expenditures from $50 billion to $70 billion in a four-year period, as consumption rose from $325 billion to $366 billion, investment rose from $75 to $86 billion, and non security expenditures rose from $50 to $52 billion. Viewed dynamically, substantial increases in the diversion process seem even less dangerous to our economy than in the static analysis, although even there they were seen to be marginal.

Much has been written recently about the degree of dynamism in the American market system as compared with the Soviet directed economy. The most reliable estimates of Soviet rates of growth in total product and industrial output seem to be about double those of the United States: total product at about 7 per cent per year, and industrial output at about 10 per cent per year, compared with American rates of 3.5 and 4.5 per cent per year respectively. The differences will probably narrow as Soviet development continues, but it may very well be that her directed economy will offer permanent differentials in growth that will favor the Soviet Union. This should not be surprising, since her whole decision-making machinery is constructed to place the highest social priority upon growth, which our market system does not do. Economists have been among the worst offenders in predicting that the Soviet system could not operate because its price system was artificial, or because its totalitarian structure interfered with incentives, or because of any of a number of reasons springing from its non-market nature. We shall have matured in the ways of national strategy when we realize that most of our preference for the market system springs from its spiritual advantages: its desirable emphasis on the individual and his importance, its scope for individual initiative, its minimal interference with individual liberty. These values far transcend, in the judgment of our society, mere economic values, but they should not blind us to the possibility that other systems stressing those economic values at the expense of the spiritual may be able to excel the market in some, or even many, economic respects.

Even granted that greater growth rates will characterize the Soviet economy indefinitely, our security problems will not be rendered hopeless or unmanageable. The size of product we now have, and which we will have in the foreseeable future, will be more than ample for our needs. The primary danger lies in the possibility that a free-market society's strong individualistic bias will prevent the necessary diversion of resources from being made. This will be the pre-eminent economic challenge to our society in the period of cold war: it will not occur in the area of product sufficiency. Reasonable flexibility in our social judgments, to allow us to adjust the priorities of collective versus individualistic objectives, must exist if we are to have a military posture sufficient to implement the determined national strategy that the Communist threat demands of us.

3

Economics and Alternative Security

Introduction

Social institutions are artificial constructs created to serve human purposes, and the purpose of the military its only legitimate purpose is to provide security by ensuring peace and protecting a way of life. Most of us believe that our own military does just that, and for this reason we have lavished resources and attention on our military forces in a spectacularly successful effort to make them ever stronger and more destructive.

Paradoxically, however, our very success has propelled us into a world in which our peace and security even our survival is problematic. Put simply, I believe that military force and the threat of military force, as a primary means of ensuring the peace and security all of us so deeply desire, is an idea whose time has gone. This is particularly so where nuclear weapons are involved. War is too destructive and the preparation for war too expensive for human society to continue to rely on the brute force of large militaries to secure its safety. Since the dawn of the nuclear age, the handwriting has been on the wall: Either we put an end to militarism and war or militarism and war will most assuredly put an end to us. Those who continue to rely on nuclear deterrence and other threats of mass violence as if this were not so are utopianists dreamers in the worst sense. They live in a fantasy world.

Because war-threatening military force is a social institution that no longer serves its purpose, it is the essence of realism to look for other, more effective social arrangements. There is nothing more realistic or pragmatic than discarding that which no longer works in favor of that which does. To be sure, it is easy to be cynical about the prospect of finding a viable alternative to nuclear deterrence or to war itself; nothing seems so immutable as the status quo. And yet the one constant in human history is change. Once slavery was a generally accepted social institution, with its own set of vested interests, deeply embedded in the status quo. Today, there is not a single nation on earth in which it is legal for one human being to own another. The very idea seems absurd. So can it be with militarism, nuclear deterrence, and even war itself.

Of course, slavery did not disappear on its own. It ended for many reasons: changing attitudes, changing technology, and changing political and economic conditions. But it ended also because there were people who believed that human beings could do better, and they were willing to combine their vision of a world without slavery with the hard work of building a practical path to their dream.

Like my co-authors in this volume, I believe human beings can do better when it comes to ensuring our national and international security. The "Machines of Armageddon" are not and need not be our sole recourse. In this chapter, I probe the ways in which economics can help in the search for more effective, alternative approaches to national and international security.

Economic Relationships and Conflict

There has long been disagreement as to the connection between economic relationships and violent conflict. The debate goes back to the Eighteenth Century arguments between the mercantilists and Adam Smith and carries through the arguments between Marxists and economic liberals today. Some (e.g., mercantilists) have argued that threats of violent coercion and the use of force are effective means of pursuing economic benefit and national power. Others (e.g., Marxists) have argued that the forceful expansion of economic activity abroad, to the extent of armed conquest if necessary, is critical to relieving social pressures at home (created by unjust economic systems) that otherwise would erupt into full-blown economic and political crises. Thus, according to these schools of thought, economic factors control of valuable natural resources, access to markets, or simply the desire to take valuable goods and services from others have been major driving forces in generating conflict and war.

Beyond this, economic relationships can create conflicts that can and do grow to the point of explosion, erupting in civil, revolutionary, and international wars. These wars often shift economic relationships in such a manner as to give rise to other conflicts that sow the seeds of future wars. Those disadvantaged at the end of the last war sooner or later seek to gain the upper hand.

It also has been argued, however, that war and preparation for war tend to damage rather than benefit national economies, especially in the long run. Writing in 1776, Adam Smith contended that expenditures on such activities were economically unproductive. I have argued that sustained high levels of military spending tend to undermine national economies by diverting critical economic resources needed for the maintenance and improvement of a nation's productive capacity.

Furthermore, economic liberals generally believe that widened international economic relationships tend to create greater opportunities for mutual benefit through cooperation rather than confrontation. Thus international economic relations may well tend to reduce the likelihood of war. In support of this position, Richard Rosecrance has written:

While trading states try to improve their position and their own domestic adocation of resources, they do so within a context of accepted interdependence. They prefer a situation which provides for specialization and division of labor among nations. One nation's attempt to improve its own access to products and resources, therefore, does not conflict with another state's attempt to do the same. The incentive to wage war is absent in such a system for war disrupts trade.

A web of economic relationships binds the participants together. Their growing interdependence does not prevent conflicts, but it does create strong incentives to settle them amicably. Furthermore, ongoing economic relationships typically imply continuing personal interaction and contact, and this tends to break down prejudices, which lower the threshold of violent conflict. From this perspective, the strengthening and expansion of the global economy may be the single strongest force for world peace today.

Which of these positions is correct? Whether international economic relationships tend to raise or lower the probability of war depends on their nature, not just their extent. Broadly speaking, exploitative relationships those in which the flow of benefit is overwhelmingly in one direction tend to increase the number and severity of conflicts whereas mutually beneficial relationships those in which the flow of benefit is more or less balanced tend to reduce the likelihood and intensity of conflict.

The reason is straightforward. Exploitative relationships are inherently unfair. Even if the parties being exploited gain something from the relationship, the fact that the vast majority of benefit flows in the other direction is bound to create or aggravate antagonisms. In either case, the exploited stand to lose little and may even gain if the relationship is destroyed. They therefore may be inclined to raise the intensity of whatever economic or other conflicts might occur, even to the point of war.

Balanced relationships tend to have the opposite effect. All parties pin from the ongoing connection they have established. Out of pure self-interest, they do not want to see the relationship disrupted. Furthermore, balanced relationships permit, even foster, the full development of every party involved. As the economic condition of each participant improves, their producers become more productive and their potential as a market grows. Thus, their capacity to contribute to the relationship increases, both as a source of products and profits.

It is a great mistake to view international economic relationships as though there always is a fixed pie of benefit to be divided, the larger pin of one party coming only at the expense of another. Exploitative relationships do tend to have this character, but balanced relationships not only provide for mutual pin, they tend also to cause the pie of benefit to grow as time goes by. As a result, when relationships are balanced, the incentives are strong to settle conflicts amicably. And as they are successfully resolved time after time, the idea of allowing them to fester to the point of confrontation comes to seem more and more absurd. Under such circumstances, the thought of brandishing the threat of war slowly recedes, and war itself becomes ultimately unthinkable. The European Economic Community is an interesting case in point.

By 1988, EEC membership included Belgium, Denmark, France, Germany, Greece, Ireland, Italy, Luxembourg, the Netherlands, Portugal Spain, and the United Kingdom, a countries that have fought countless wars with each other over the centuries. Yet, today, if you were to ask their citizens what they thought the prospect of war between their homelands would be over the next fifty years, it would not be considered a serious question. Quite the contrary, trade, travel and cooperation among their nations is growing; the bonds are strengthening as they move to eliminate all remaining trade restrictions by 1992 and contemplate establishing a common currency.

True, the military cooperation of many of these nations in NATO doubtless has played some part in diminishing the possibility of war among them. Yet membership in NATO not in the EEC did not stop Greece and Turkey from facing off against each other militarily during the Cyprus crisis of the mid-1970s. Furthermore, it is not uncommon for military alliances to dissolve quickly into serious military confrontation or war. The US-USSR shift from military allies to antagonists after World War II is a striking example. A comparable disintegration of the Common Market would be astonishing.

It is easy to see why someone being exploited in an unbalanced relationship would be better off if the relationship were to become more balanced. But the dominant party, over the long term, would gain from greater balance in most cases also. A dominating party in an exploitative relationship generally has to expend considerable effort to maintain control and this effort is much more of a drain than is commonly supposed. When, however, a relationship is balanced, there is no need to expend extra effort to keep it going. The mutual flow of benefits binds the parties together. Thus, a balanced relationship is in effect a more efficient relationship; the benefits are achieved at much lower cost.

Of course, in any exploitative relationship the expense of maintaining control potentially can be reduced if the exploiters are able psychologically to manipulate the exploited to make them feel helpless or accepting even deserving of their subordinate position. This is the basis of what may be one of Karl Marx most important insights, the concept of "false consciousness" a concept that can be extended to an even more effective method of maintaining control, namely, getting people to identify with the very system that is exploiting them. If appeals to patriotism or to the grandeur of empire succeed in making people proud to be part of such a powerful institution, they not only may accept but actively support a system that is in fact exploiting them.

But even false consciousness does not work forever; and though the benefit to the dominator in an exploitative relationship may be greater than if the relationship were balanced, even after allowing for the cost of maintaining control, that tends to be true only in the short run. In the long run, continued exploitation breeds discontent and disaffection, the exploited come more and more to see the possibility and desirability of change, and the dominator in the exploitative relationship consequently suffers heightened insecurity and expense in maintaining control. As Adam Smith concluded more than two hundred years ago, after a lengthy discussion in the context of the British colonial empire:

Under the present system of management, Great Britain derives nothing but loss from the dominion which she assumes over her colonies.

Great Britain should voluntarily give up all authority over her colonies. She would not only be immediately freed from the whole annual expense of the peace establishment of the colonies, but might settle with them such a treaty of commerce as would effectually secure to her a free trade, more advantageous to the great body of the people than the monopoly which she at present enjoys.

In sum, exploitation tends ultimately to become counter productive. But if it is true that exploitation is counter productive in the long run even for the exploiter, why do we continue to believe that we are best off if we are dominant in a relationship? There are, I think, two main reasons.

The first and most obvious reason is myopia. Human beings can much more easily calculate and relate to the short-term effects of their actions than the long-run consequences. Even if a certain behavior is known to be counterproductive in the long run, it is very difficult to get people (or their nations) to stop if the short-term effects are thought desirable. Cigarette smoking and chronic overeating (or over dieting) come easily to mind as examples for individuals.

The second reason may well be yet more compelling. We come to know, virtually from birth, that for much of what we need or want in life we must depend on others. From infancy, our growth is largely a transition from total dependency to greater self-reliance. Yet even in adulthood, we remain intellectually, emotionally, and physically dependent on others for guidance, interaction, praise, and support, as well as for the material goods and services we want and need. This creates a degree of insecurity that comes from the fear that those on whom we must depend may not do what we want or need them to do. To some extent, we have learned to deal with our insecurity through a paradigm of domination. If we are in a position of control, we can manipulate anyone who will not go along voluntarily to give us what we want or need.

There is much in the human experience, as individuals and as nations, to support the idea that those who dominate often live at a higher material standard of living, at least in the short run. It is much harder to see the negative long-run consequences for the exploiters or that their dominance does not necessarily translate into a higher quality of life even in the short run that is, when all of life's critical intangibles (such as love, self-respect, and feelings of fulfillment) are taken into account.

The paradigm of domination may never have been optimal in international affairs, but it has become increasingly dangerous, for the machines and institutions of nuclear war are the logical and inevitable creations of this paradigm. They are the ultimate coercive threat for the purpose of dominating others and the ultimate counterthreat to frustrate the attempts of others to dominate us. This system of threat and counterthreat has lead us inexorably closer to the brink of the war we all dread. It has become utterly self defeating.

The threat or use of force is actually a much less effective way of getting others to do what you want them to do than taking a more positive approach for example, by making friends or creating working partnerships. It is an obvious lesson of everyday life that we fail to apply to the realm of international relations because our thinking has been so narrowed by the paradigm of domination. Suppose, for example, that some neighbors have been playing their stereo very loudly at night and that it really disturbs you. You could pound on their door and threaten to have them arrested if they do not stop the noise. It is possible they will stop, but it is even more likely that they will get angry and either make the music louder or find some other way of annoying you. Suppose instead you invited your neighbors over for coffee and took the time to try to establish a more friendly relationship. It is possible they would continue to play the music loudly, but it is much more likely that they would make the music softer. In fact, they are likely to turn the volume down as soon as they become aware that the music is bothering you. And to the extent that you do become friends, they generally win try to avoid creating disturbances in the future.

There are circumstances in which coercion may be required. But the establishment of friendly relations is much more likely to be effective than threatening to use or actually using force. The use of military deterrence (nuclear or otherwise) is inherently threat-based, coercive, and unbalanced. It is an inferior way of creating security or of achieving other national objectives. If it is possible to strengthen positive incentives for the peaceful resolution of conflicts, that is very likely to be a less costly and more effective path to increased security and greater economic well-being.

When is it possible to replace more distant or hostile relationships with friendlier ones and how might that be done? Across the spectrum of relationships from the personal to the political these are not easy questions to answer. Clearly, willingness and commitment to undertake initiatives aimed at breaking down barriers play a key role. But factors internal to the parties may also be critical in increasing receptiveness to friendlier ties. Both of these considerations were obviously at work in dramatically improving relationships between the United States and the People's Republic of China (PRC) in the 1970s and between the United States and the Soviet Union in the late 1980s. Political and economic changes within the PRC, in part resulting from internal economic problems, greatly increased the willingness of the Chinese to respond to initiatives from the United States. Similar pressures and changes in the USSR pushed the Soviets to make a startling series of initiatives to establish friendlier relations with the United States, even in the face of an extremely hostile US administration -and these initiatives clearly succeeded.

Strategies for a Peacekeeping Economy

In his intriguing book Stable Peace, Kenneth Boulding sets forth what he calls the "chalk theory" of war and peace: a piece of chalk breaks when the strain applied to it is greater than its strength, i.e., its ability to resist that strain. Similarly, war breaks out when the strain applied to an international system exceeds the ability of that system to withstand strain. The establishment of "stable peace" thus requires that strains be reduced, that strength be increased, or both.

In maximizing the peacekeeping potential of the international economic system, it seems most fruitful to look for a combination of strain-reducing and strength-enhancing strategies. The four basic strategies that follow are, I believe, key to moving toward a peacekeeping international economy.

Strategy I: Balance Independence and Interdependence

Increasing independence tends to reduce strain because it reduces vulnerability to externally generated disruptions of the nation's economy. Vulnerability tends to create insecurity for a number of reasons. There is the fear that opponents will purposely exploit vulnerability to pin the upper hand in a conflict (e.g., by cutting off the flow of key goods or resources) or that others will unintentionally harm you while pursuing their own objectives (e.g., by inventing substitutes for the exports on which your economy depends). These fears may be exaggerated or they may be justified, but in either case the insecurity they generate typically produces defensive, belligerent behavior. A greater degree of independence helps to avoid these problems.

Increasing mutual interdependence tends to increase strength because it increases the incentive to avoid disruption. The web of interdependence created by balanced relationships yields a flow of mutual benefit that will be reduced or eliminated if the conflicts that arise are permitted to get out of control. This cannot help but encourage the peaceful resolution of conflict, increasing the resistance of the international system to strain.

How is it possible to create an economic system whose actors are both independent and interdependent? One approach is to reduce dependence in those areas of interaction in which vulnerability is most frightening while increasing it in other areas where there is a real potential for mutual gain. There is no doubt that dependence on outside sources of supply is more troubling the more critical the good or service in question. A nation is more exposed and vulnerable if it must rely on foreign suppliers for critical elements of its food supply than if it is a large net importer of television programs.

Now it might be argued that a nation that is dependent on an outside supplier of critical goods would be even less likely to go to war with that outside supplier than if it were independent. But unless the critical goods dependence is mutual the nation supplying the critical goods has no comparable incentive to avoid war or other coercion against its customer. And, as argued earlier, it might even feel encouraged to behave coercively because of the leverage the criticality of the goods provides. The nations of the Organization of Petroleum Exporting Countries (OPEC), acting jointly, gained considerable economic and political benefit by using their critical goods leverage in the 1970s. Of course, the possibility that critical goods suppliers might try to use their leverage in this way gives strong incentives to the dependent nations to try to keep the supplying nations under control. Over time, a range of coercive policies from mild to severe have been used for this purpose, from manipulating corrupt officials to forced colonization.

One nation's dependence on another for critical goods might encourage a third party to behave more aggressively in dealing with the dependent nation when conflicts erupt. Why? Because the third party may be confident that it can disrupt the flow of critical goods and thereby quickly gain the upper hand, particularly if the critical goods supplier is itself a weak nation located far from its customer. This was part of the Japanese strategy in World War II. For all these reasons, the goal of independence in essential goods seems worthwhile.

Each nation's attempt to balance its independence and interdependence flows from its attempt to balance the political gains of reduced vulnerability against the economic gains of trade. It is not necessary to assume that nations act out of an unselfish desire to do their part in maximizing the economic well-being of the world as a whole. The pursuit of more narrowly defined self-interest is sufficient.

Which goods and services are most critical to a nation's economic well-being depends in large part on the nature of its economy and the level of its development. Reliable supplies of steel are far more critical to an industrialized nation than to one that is mainly agricultural; reliable supplies of tractors and harvesting machines are more critical to a highly developed agricultural nation than to either a nation that is largely industrial or one that is agricultural but much less developed. There are, however, some categories of goods that may be broadly considered critical.

Foods that have few substitutes and that play a central role in the diet of the population are clearly in the category of critical products. Thus it is desirable that each nation be capable of meeting the basic nutritional requirements of its own population. Similarly, potable water is critical to health and well-being. This fact not only implies that each nation should be able to provide minimum water requirements without depending on foreign suppliers but also that nations should be capable of ensuring the purity of that water. Energy supplies are critical to any economy, though they tend to be more completely integrated into those economies that are more highly developed. And every economy requires continuing supplies of certain key raw materials. Thus, independence in terms of energy and critical raw materials is desirable as well.

It is not necessary for each nation to be totally self-sufficient in critical goods. To avoid the insecurity inherent in dependence on an external lifeline of critical goods, a nation need only be able to supply the minimum amount of those goods it requires from internal sources during situations of prolonged disruption of trade. For example, a nation need not be able to produce locally its entire supply of food; it need only be able to supply sufficient food to meet the basic nutritional requirements of its population if and when it is cut off from foreign sources.

Nations unable to achieve independent domestic production of even minimum quantities of critical goods may still be able to achieve a degree of independence in domestic supply. When the critical goods are storable, it may be feasible to establish national stockpiles that provide a significant short-term buffer. That would at least buy time for a more reasoned response in the event of a sudden and arbitrary disruption of supply. If the buffer were sufficiently large, enough time would be available to pursue a number of avenues for the non-violent resolution of the conflict that brought about the disruption. It also would allow time for arranging alternative sources of supply. A large buffer might even serve as a deterrent to punitive disruption because the nation that contemplates interrupting the supply would know in advance that its action would be relatively ineffective. Similar results could be achieved also by developing a standby domestic production capability for critical goods.

Switzerland has adopted a combined standby domestic production capability and "strategic stockpiles" approach as part of its comprehensive national defense strategy. As Dietrich Fischer has written:

During peacetime, Switzerland imports nearly 50% of its food consumption. In case of a cutoff of imports, food would be immediately rationed, requiring about a one-third reduction in daily caloric intake. Meat consumption would be drastically reduced to arrive at a more efficient calorie conversion factor through a more vegetarian diet. Grassland would be converted gradually, in three yearly phases, into cropland. In the meantime, the food deficit would be bridged with reserves of non-perishable food, which are constantly being renewed in peacetime. Similar programs exist for fuel certain minerals and other vital commodities.

On the other hand, interdependence should be maximized with respect to non-critical items. There are many goods that are non critical yet very much in demand. Examples include recorded music, movies, foods that add pleasant variety to the local diet, books, more stylish items of clothing, sports equipment, cosmetics, perfume for that matter, nearly all goods in which there is a flourishing trade in peaceful times. For interdependence to play its role in increasing the strength of the international system most effectively, it cannot be restricted only to goods that are of marginal interest to the economies involved. A flourishing trade in trivial goods win do little to bind nations together because the mutual benefits of such trade are minimal. Essentially then, a widening and deepening of international trade should be encouraged, with the exception of a relatively restricted set of especially critical goods.

The economic theory of comparative advantage holds that the world total of goods and services available is greatest if each nation specializes in those products it is best at producing, then trades for those goods and services it does not produce. It makes sense that a system in which everyone concentrates on what they do best and depends on others for the rest of what they need is potentially more efficient than a system in which everyone tries to be self-sufficient. Specialization and trade is in fact key to the internal modus operandi of modern economies.

The strategies recommended here preserve the spirit of this theory. But there are two significant departures. First, critical goods are exempted for political and psychological reasons. Second, the theory of comparative advantage is really more appropriate to a static than a dynamic view of the world. Developing the full economic potential of a nation inevitably will change the structure of its economy in ways that are likely to alter the pattern of comparative advantage. It is crucial to make room for such change. To require that each nation remain frozen in its present pattern of relative advantage is to hinder, if not to checkmate, its prospects for economic development. Establishing a peacekeeping international economy requires encouraging, not frustrating, the process of economic development. Moreover, it is best if the economic interdependence of nations involves multiple sources of supply and multiple markets. A web of multilateral dependence will tie more nations more closely together than will a set of disconnected bilateral relationships. The continued economic health of each nation win come to depend not only on the well-being of its own trading partners but also on the welfare of those nations that trade with its trading partners. Broader and stronger incentives for maintaining the peace in the face of conflict will therefore be created.

Of course, though generally it is wise to emphasize positive economic incentives, situations may arise in which negative economic sanctions (e.g., boycotts) may be useful (e.g., in instances of gross violations of human rights); and the question therefore arises: Would a web of multilateral dependence make such sanctions less effective? It is true that a web of multilateral dependence would render each nation less vulnerable to sanctions imposed by any other single nation. But widely supported and concerted multilateral sanctions can still be effective. Indeed, when sanctions are applied multilaterally, they are likely to be more powerful economically, politically, and psychologically. Moreover, the need to convince other nations to go along will help to prevent sanctions from being applied too casually and too often.

Strategy II: Balance Relationships

For reasons discussed earlier, maximizing the peacekeeping capacity of the international economy requires creating a situation in which the flow of benefits in every international trade relationship is as balanced as possible. Perfect bilateral balance is not required. It only is necessary that enough benefit flows in both directions to create the perception that neither side is being exploited by the other, and that continuation of the relationship on more or less the same terms is clearly in the interest of both parties.

Economists typically assume that an trade transactions are voluntary and that all parties are rational maximizers of their own well-being. Therefore, the very existence of trade between nations demonstrates mutual gain, because rational traders would not voluntarily exchange anything unless what they got was worth more to them than what they gave up. But these assumptions do not imply that gains will necessarily be balanced.

For example, if a person buys a piece of land for $70,000, an economist would assume that both the buyer and seller gain from the exchange or they would not have agreed to it. But it is perfectly possible that the land is worth $60,000 to the seller (who therefore gains $10,000 from the trade) and $2,000,000 to the buyer (who therefore gains $1,930,000). Furthermore, even though both have gained, if the seller later learns that the buyer would have been willing to pay almost $2,000,000 for that land, it is likely that the seller will feel angry and exploited, poisoning future relations between them. If the transaction had taken place at a price closer to $1,030,000, on the other hand, both buyer and seller would have gained somewhere around $970,000 from their trade. Gain would be balanced and future revelations would be less likely to cause either party to feel angry or exploited (which of course is not to say that the buyer would not have preferred to pay the lower price).

Differences in market power, information, and bargaining ability can easily result in an unbalanced flow of benefits. It would be useful if there were some way of objectively establishing the degree of balance in trade relations to guide this strategy. There has been some attempt to develop a sensible objective standard, but there is still much that remains to be done.

Balanced gain, however, is only one dimension of balanced relationships. It also is important that input into the decision making process as well as decisional power itself be balanced. If the decision process relevant to the relationship is lopsided, one side may feel that it is excessively dependent on the good graces of the other a situation not conducive to feeling secure. On the other hand, a relationship in which both parties are equal partners in decision-making gives each of them the feeling of ownership in the relationship. It is theirs, something that they have created, not simply a gift that one of them unilaterally bestowed on the other and can just as easily withdraw. Both parties therefore tend to feel a responsibility for the "care and feeding" of the relationship, an obligation to help it succeed and to continue. Such feelings cannot help but strengthen the system.

Whether at the level of international treaty negotiation or interpersonal conflict mediation, imposed solutions no matter how clever are rarely as effective in dispute resolution as solutions to which both parties have contributed. If the parties themselves have participated in the creation of the agreed solution, they will tend to have a better understanding of the underlying issues and more confidence in the fairness of the settlement. Even if neither party is thrilled with the final result, there will be a greater tendency for each to abide by it. Also, and just as important, they will be more confident that the process of mutual interaction that brought the conflict to a successful resolution is capable of resolving future conflicts as well.

Strategy III: Create Flexible, Development-Oriented International Financial Institutions

Financing mechanisms can reduce strain and increase strength or add to strain and reduce strength. Properly structured financial mechanisms transfer present claims on economic resources (in the form of money) from those who have accumulated them to those who can put them to productive use. The smooth functioning of financial institutions thus prevents situations in which those with excess funds to invest cannot find fruitful uses for those funds, while those who have profitable opportunities for investment are not able to pin access to sufficient funds to take advantage of them. However, financial mechanisms can be used also as a tool for domination and control. Readily available credit on terms that make it easy to borrow is very seductive, particularly to those who can least afford it. An individual, company, or nation that gets itself too deeply in debt will soon discover that it has compromised, if not completely lost, its independence. The colossal expansion of international debt in recent decades has left the sovereignty of more than a few nations in question. This is particularly (although not exclusively) true of the world's less developed nations. By 1985, the external debt of the Third World was greater than $750 billion, more than 330 percent higher than it was in 1975. As these nations have fallen more deeply into debt, they have been forced to make very difficult and painful choices.

The rising cost of servicing the debt began to absorb much of hard currency export earnings of these Third World nations, limiting their capacity to import both investment goods needed for development and consumer goods needed to maintain present living standards. To the extent that this interfered with the development process, the very growth of debt actually reduced their ability to generate the income needed to continue servicing the debt. To default on their loans would have created considerable international ill will as well as making it extraordinarily difficult to borrow in the future. Some Third World debtors had to borrow still more money just to make the required payments on their present loans. But given their increasingly fragile economic condition, international bankers were unwilling to lend additional funds to them unless they agreed to such draconian measures as wage freezes, devaluation of domestic currencies, and removal of government subsidies on food and other vital goods. Although such measures may be viewed as critical correctives by orthodox conservative banking principles, they deliver a severe economic shock to the local population, sending the price of many necessities soaring at the same time they prevent wages from rising. In more than a few cases the result has been considerable domestic unrest, even revolution. For the government of any nation to be put into the position of having major domestic policy measures dictated by foreigners is a bitter pill to swallow. In the case of former colonies, this compromise of national sovereignty has led, not surprisingly, to cries of "recolonization." Whether this situation is the result of conscious attempts at domination or the best of intentions gone wrong, whether it is the fault of the more developed countries, the less developed countries, or the international bankers, it has created a great deal of real hostility, conflict, and insecurity. And that is something that a peacekeeping economy must avoid. Furthermore, deepening international debt, combined with less than stellar economic performance on the part of many debtors, also has made real the prospect of widespread default. Such a possibility is a serious threat to the stability of the international banking system, and thus to the economies of many nations, both rich and poor.How, then, is it possible to construct sound international financial mechanisms that provide the economically critical lubricating function of finance while minimizing its conflict generating tendencies? There are perhaps four basic principles.

1. Maximize the extent to which debt financing is targeted to projects capable of generating the additional wealth needed to pay off the debt. Many nations have borrowed enormous sums of money internationally to finance a variety of expenditures that have no prospect of generating additional economic wealth. This of course means that repayment of the debt causes a net reduction in economic well-being, as existing wealth flows out of the country.

   Debt-financed military expansion is one of the most common examples. In the early 1960s, the less developed countries as a group accounted for about 10 percent of world military expenditures. By the early 1980s, they accounted for more than twice that large a share. If the Third World's share of world military spending had been the same over the first six years of the 1980s (i.e., 1980-1985) as it was in the early 1960s, the savings from this one source alone would have amounted to more than half of their outstanding 1985 debt.
2. Tie the fortunes of the providers of the capital to the success of those particular projects. As long as the borrower (or others

who stand ready to pay off the debt should the borrower default) has accessible sources of wealth sufficient to pay a loan, the lender has little incentive to worry about the viability of the particular project being financed. It is better to strengthen the incentives of the providers of capital to see that the project at hand succeeds. Joint ventures are one effective means of accomplishing this.

3. Limit the burden of debt financing on borrowers. This will both create a greater balance of benefit and avoid conflict-generating resentment on the part of the borrower, who might otherwise come to see repayment of the loan as yet another source of exploitation. Limiting the debt burden also allows borrowers a better chance of getting out from under debt associated with unsuccessful projects.
4. Provide outright grants or heavily subsidized long-term loans for long-term public projects that are critical for development, but are not profitable in the narrow business sense. There are many projects that would fail ordinary accounting criteria of profitability that are nonetheless worthwhile even vital to the success of development because they generate crucial social benefits that are difficult or impossible to measure. Infrastructural projects such as road building, improvement of sewage treatment and water supply facilities, and construction of schools often have this character. Ordinary debt financing of projects of this sort is inappropriate by the first principle and could well be discouragingly burdensome.

Strategy IV: Minimize Ecological Stress

Much of the world's ecological stress is the result of the economic activities of production and consumption. A high level of transnational ecological stress increases strain and is, therefore, inconsistent with a peacekeeping international company. Continuing international conflicts are created by, for example, chronic problems like acid rain. Acute environmental problems such as those caused by the breakup of an oil tanker or the meltdown of a nuclear power plant periodically raise the level of international confrontation still higher. Transboundary pollution by itself may not lead to war, but there is little doubt that it already has generated a degree of conflict and hostility and that it has the potential for generating a good deal more. According to Michael Renner:

Massive damage from acid deposition in Canada (more than 50% of which comes from U.S. sources) has caused considerable diplomatic friction between Canada and the United States. The Danish parliament decided to ask Sweden to close a [nuclear power] plant 30 kilometers from Copenhagen. The French government rejected a similar plea by local West German authorities to cancel construction of four reactors. Tensions on similar issues run high between Ireland and the United Kingdom, between Austria and both West Germany and Czechoslovakia, between Hong Kong and China, and between Argentina and Chile.

Every additional source of tension contributes to the strain in the international system and thus to the likelihood that other sources of conflict will lead to the eruption of violence. The greater the load on the camel's back, the more likely the next straw will break it.

Perhaps even greater political conflicts have been generated by another form of ecological stress: the depletion of vital nonrenewable natural resources. In fact, as discussed earlier, it has been argued that the desire to gain (and often, to monopolize) access to raw materials has been a significant cause of war historically. There is little doubt that it was one of the driving forces behind the colonization of much of the world by the more economically and militarily advanced nations. And though the imperial era has passed, these pressures continue to bring nations into conflicts of the most dangerous kind conflicts in which each party believes that what is a stake is the continued economic well-being and perhaps political sovereignty or survival of its own people. It is hard to imagine that the Middle East, for example, would continue to inspire as much of the attention of the major military powers as it does today if it were not a source of critical supplies of crude oil. Such conflicts create considerable strain and thus a compelling threat to international peace. They must be minimized.

It has been argued that the expansion of economic activity itself is inconsistent with the maintenance of environmental quality, that modern activities of production and consumption inevitably generate a degree of ecological stress. There is an element of truth in this. Yet the levels of economic well-being to which populations of the more developed countries have become accustomed can be maintained, improved, and extended to the less developed nations without generating even the levels of environmental degradation we currently are experiencing. Doing so, however, requires (1) a great deal more attention to the efficient use of natural resources; (2) the development and use of pollution-abating technologies and procedures; and (3) a shift toward qualitative rather than quantitative economic growth.

Characteristic Structures and Institutions of an International Economy

There are a great many different specific structures and institutions that are consistent with the four basic strategies of a peacekeeping economy considered above. What follows is intended as illustrative rather than definitive, a first attempt to outline one feasible structure.

Generally Free Trade

A peacekeeping international economy would be characterized by relatively few barriers to trade. As long as trade relations are mutually beneficial a large volume of trade will help bind nations together. In a world of mutually beneficial trade relations, all nations that are potential trading partners will tend to look upon tariff barriers and other such obstacles as a nuisance. The primary exception to the pattern of free trade would lie in the area of critical goods.

Tariffs and other trade barriers may be useful tools for encouraging greater national independence in critical goods, especially in the short run. If it is possible to arrive at general international agreement concerning what goods may be considered critical, the erection of barriers to trade in such goods will be understood as limited and will not result in a widening pattern of retaliatory tariff actions by other nations. However, although very useful, general consensus on what constitutes a critical good is not crucial. Failure to come to such agreement would simply leave each nation to concoct its own operating definition and to deal individually with any antagonism its definition might create among its present or potential trading partners. This would be less than ideal but stiff workable.

Other approaches to achieving independent critical goods supply capability may be less contentious than encouraging domestic production behind a protective wall of tariffs and other trade barriers. Direct government subsidy of domestic critical goods production may be cheaper, and less likely to encourage retaliation. Contingency plans for emergency production of critical goods, as in the Swiss example cited earlier, are even less likely to cause problems. They do not interfere at all with trade under normal conditions. Strategic stockpiling of storable goods is even less likely to generate international hostilities because it actually encourages the purchase of such goods, especially during the initial buildup of stores.

The widespread use of critical goods trade barriers might seem to make life especially difficult for those nations whose main exports are critical goods. But it is highly unlikely that all nations will be able to achieve complete independence in critical goods, at least in the short term. There is very little chance that a country like Japan, for example, will achieve independence in critical raw materials or energy in the foreseeable future, given that nation's relatively poor endowment of depletable natural resources. The contingency plan and stockpiling options are still open to such countries, but they would not interfere with the sales of critical goods exporters.

Furthermore, as it is necessary for any nation to be capable only of meeting its minimum critical needs independently during prolonged trade disruptions, there is plenty of room for trading internationally in normal times. Such trade is encouraged by the desire to increase the quantity, quality, and variety of critical goods available. Food is a particularly clear example. A nation may work to develop its capability to meet the minimum nutritional requirements of its population from local grain production or from stockpiles during emergencies, but under normal conditions access to imported fruits, meats, vegetables, and processed food products win make its population's diet a great deal more pleasant and interesting.

In general, the peacekeeping economy would be characterized by relatively free international trade in which multiple sources of supply would be the norm. Multiple sources are especially important for critical goods, but should be the basic structure of all trade.

One way to achieve such a trade pattern would be through a combination of regionally integrated international free trade zones, like the European Economic Community (EEC) and broader interregional trade arrangements. The establishment of regional common markets has two potential advantages. First, it is smaller and easier to manage than a quantum leap to a truly integrated, free-flowing world marketplace; yet at the same time it is consistent with movement in that direction. It is thus a logical interim step. A pattern of gradually widening regional common markets is a reasonable, evolutionary path to global free trade. Second, to the extent that they are successful regional economic communities create a base of cooperative experience and mutual gain that can lead easily to a higher degree of political as well as economic cooperation. The success of the EEC has led to the establishment of a fledgling European Parliament and a plan to eliminate virtually all remaining intra-regional trade restrictions by 1992. It also recently has given new credence to proposals for a single European Central Bank and a unified currency for Western Europe-proposals with profound political as well as economic implications.

Greater Trade and Economic Cooperation Among the Less Developed Countries

It is easiest to establish balanced relationships between parties of more or less equal power. Accordingly, greater cooperation among the less developed countries (LDCS) can help to break down the colonial mentality and patterns of dependency that are the legacy of the subordinate role established in the age of empire. Of course, it is misleading to suggest that all LDCs are at a comparable level of economic and political development or even that there is necessarily a great deal of homogeneity among them. However, LDCs are on the average closer to each other than to the more developed countries (MDCs). And for the most part the paradigm of dominance is not as historically entrenched in their interactions as it is in LDC-MDC relations. For both those reasons, it should be easier to establish mutually beneficial, balanced LDC relationships compatible with a peacekeeping economy.

A number of regional LDC free-trade zones have been established, with varying degrees of success. The Central American Common Market (CACM) was formed by five nations in 1960 and disbanded as a result of political differences in the 1970s. Fifteen countries with a combined population of about 124 million established the Economic Community of West African States (ECOWAS) in 1975. It continues to function today, but it cannot be said to have been an unqualified success.

Of course, greater trade among the LDCs does not depend upon the formation of regional free-trade zones. There are good economic reasons why a general broadening of trade among the LDCs seems worthwhile. Third World producers should better understand the context and constraints of societies at lower levels of development than do their MDC counterparts. They should therefore be more attuned to manufacturing products whose design features, economic characteristics, and operating and maintenance requirements are more appropriate to LDC customers.

There are supply-side advantages as well. LDC manufacturers who strive to service MDC markets must meet more rigorous MDC standards for product quality. But quality is not cost free. Insisting on products of high enough quality to satisfy wealthier MDC customers typically will require a production process so expensive that it will price those products completely out of reach of all but the elites in the LDCs. Cheaper modes of production are likely not only to be more appropriate for LDC producers but also to produce goods that can be sold at prices that will make them accessible to a much larger fraction of LDC populations.

There are potential advantages to expanded LDC cooperation in marketing primary products such as food and raw materials to the MDCs. The success of price actions begun by the Organization of Petroleum Exporting Countries (OPEC) in 1973 is a spectacular example of the ability of producer coordination among LDCs to achieve prices for their primary products that more closely reflect their value to MDC customers. It would have been wiser for the OPEC nations to have raised prices more slowly to avoid the sudden shock and consternation their actions created. And it is a great pity that the gains to the OPEC nations were not used to mitigate the oil price shock on other LDCs and for more developmentally oriented purposes at home. Nevertheless, OPEC's success in shifting the international flow of benefit was impressive.

It may seem odd to argue that more monopolistic behavior on the part of the LDCs might be a good idea. After all, one of the central points of this analysis is that a peacekeeping economy must move away from the kind of exploitative behavior for which monopolies are famous. But the LDC producers in these markets often find themselves facing off against powerful and sophisticated MDC governments or multinational firms with which they have very little chance of striking an equitable bargain. Such a market, dominated by powerful monopolistic forces on one side, can move closer to a more equitable and efficient competitive outcome either by breaking up those forces or by increasing the power of other players in the market to countervail that power. John Kenneth Galbraith has suggested that there might be circumstances in which the latter approach could have advantages over the more traditional antitrust actions.

This is clearly not an ideal situation. There is, for example, the very considerable danger that powerful LDC cartels will cooperate with powerful MDC-based multinational corporations rather than facing off against them, to the detriment of both MDC and LDC consumers. Worse yet, LDC cartels that abruptly force prices as high as did OPEC in the mid-1970s may create so much hostility as to increase strain in the international system. Therefore, it is worth giving careful attention to the search for other practical mechanisms capable of better balancing the international flow of benefits without resorting to the cartelization of LDC primary product exporters.

Special Purpose International Funds

There are three special purpose international funds that would be useful in achieving or at least facilitating the objectives of a peacekeeping international economy: a Global Environment Fund (GEF), a Global Infrastructure Fund (GIF), and a Global Bank.

Global Environment Fund

There are cases in which the consumption or production activities of one nation generate pollution that causes ecological damage abroad. Acid rain is a case in point. Suppose Country A is creating pollution that results in acid rain falling on Country B. Substantial improvement in environmental quality in Country B can be achieved if Country A either halts the pollution-generating activity or institutes antipollution measures. If Country A ceases that activity, it loses the benefits associated with it; if it cleans up the activity, it bears the costs of the cleanup; and, in either case, the main environmental advantages accrue to Country B. Operating out of narrow, short-term economic self-interest, Country A has no incentive to cease polluting the environment of Country B. Now, suppose Country A is an LDC whose resources are already stretched to the limit in its attempt to meet its own needs and to get sustainable development going, while Country B is an MDC in a relatively strong economic and technical situation. Would it not make sense for Country B to offer assistance to Country A to help in the abatement of this pollution?

A Global Environment Fund could provide grants to LDCs to assist them in undertaking projects with strongly positive environmental impacts. These might involve improved sewage treatment facilities, air pollution control measures, protection of endangered species, or preservation of rare environmental assets such as rapidly disappearing hardwood forests. Where it is ideologically acceptable, the GEF itself might buy land in a given country and hold it as an environmental preserve. Or it might provide a "perpetually forgivable" loan to the government or to an indigenous group to buy the land and preserve it. Loan payments would be due to begin at a specified time each year, but at that time the due date would be postponed another year if the area still remained untouched. As long as the area purchased remained an environmental preserve, no loan payments ever need come due. Thus, the Global Environment Fund would help to institutionalize Strategy IV of a peacekeeping international economy, the minimization of global ecological stress.

Global Infrastructure Fund

The GIF would be similar to the GEF in its structure and function but would finance projects to improve Third World systems of transportation, communication, and water and energy supply. Such a fund has been proposed by Masaki Nakajima, an economist who headed Japan's Mitsubishi Bank. The Nakajima "global Marshall Plan" would be financed by the MDCs, making available $25 billion per year as seed money for mega-projects "that would have worldwide impact but are too costly to be undertaken by a single nation, such as massive water projects to 'green' the world's deserts." The problem is, however, that the economic impact of such "macro-engineering" is likely to be more positive for the MDC construction firms that build them than for the Third World. Furthermore, the environmental impact of such grand projects is questionable.

What is needed instead is a grant-making fund that finances a variety of infrastructural projects, many of them small or moderate in scale. For example, the digging of many individual wells may be a more effective and environmentally sound approach to irrigation in some places then building huge dams and pipelines. Building a network of relatively simple roads may be more a contribution to development than building modern superhighways that span continents. Large-scale projects may sometimes be desirable, but the highest probability is that small- to medium-scale projects will make the most sense most of the time.

Perhaps the most critical area for the GIF is in improving the quality and availability of potable water. It has been estimated that more than three-quarters of the disease in the LDCs is traceable to unclean water. Insufficient water also threatens food supplies. Drought, along with a great deal of political disruption, has brought, for example, mass starvation in recent years to Ethiopia, Somalia, and the Sudan. Projects to address this critical need are a natural area of focus for the GIF and cooperation between the GIF and the GEF.

Thus, the GIF would be responsive to Strategy III, i.e., providing a financial mechanism less likely to generate hostility and cries of economic recolonization than present systems. Also, it would encourage real economic development, not just economic growth (see this chapter: Sustained Multidimensional Efforts in Development).

Global Bank

A Global Bank committed to providing financing and technical assistance for development projects is an extremely good idea also. The present World Bank may offer a useful starting point. Unlike the GEF and GIF, the Global Bank would not provide outright grants primarily but, rather, loans and limited-return equity financing.

Care must be taken to avoid creating an insupportable level of debt that interferes with, rather than encourages, development. Some low-interest, long-term development loans should be provided, but the Bank should emphasize equity financing for particular development projects with profit potential. In effect, the Bank would buy the equivalent of non-voting preferred stock in the project. If the project became profitable, the Bank would receive limited dividends, as is typical of preferred stock. At any time, the LDC organization involved could buy back this preferred stock at its original price. With such financing, unsuccessful projects would not create a continuing burden for the LDCs, yet successful projects would not see a disproportionate share of their return continuing to leave the country. The LDCs' downside risk consequently would be limited without unduly restricting the profit incentives of LDC investors.

In sum, the GEF, the GIF to lesser extent, and the Global Bank, each unsustainable in the absence of continuing external support, would not be profit-making institutions but, rather, mechanisms through which the economically stronger MDCs could provide aid to the LDCs on a multilateral basis. If MDC participation were broad enough, it would be possible for these institutions to reduce or avoid the more blatant political manipulation and abuse that so often has accompanied bilateral aid in the past. The primary MDC incentive to offer this aid would be that it would constitute a more effective and cheaper way of buying security than continuing to pour huge sums into oversized, insatiable militaries. Other incentives would include improved environmental quality and greater economic development with the potential for expanding markets and improving the quality of life of MDC as well as LDC populations.

Sustained Multidimensional Efforts in Development

Economic growth is simply a matter of expansion in the size of the economy. The fact that an economy is growing does not by itself assure that the material condition of the population is improving. Economic development is a complex matter of qualitative as well as quantitative change in skills, productive capacity, level of output, health, nutrition, education, organization, and general material well being. Real development raises capability, provides opportunity, and improves living standards, not just for a privileged few but for the broad mass of the population. Until there is a great deal more real economic development and not merely economic growth in the Third

World, it will be difficult for the peacekeeping economy to reach its full potential as a source of global security. It is much easier and more natural for the kind of mutually beneficial relationships that characterize the European Common Market to arise and persist among nations at similar levels of development, high enough to render them valuable to each other as customers and suppliers. If they were all comparably poor, they would have much less to offer each other, which is one of the reasons why some present and past LDC attempts at common markets have not been as successful as they might otherwise have been.

Generating sustained economic development is one of the most important elements in creating greater security through a peacekeeping economy. A rising standard of living tends to reduce the desperation people feel when they are hard pressed to provide themselves with the material necessities of a decent life. True, those living in the most extreme conditions of poverty and deprivation are unlikely to have the strength or the means to engage in sustained social violence. All of their energies are expended in a constant fight to survive. But a great many of the world's poor are not at the barest edge of survival and they have strong incentives and at least minimal means with which to struggle for change. They are much more likely to turn to extreme solutions and generally are more willing to support, even to participate in, disruptive violence than people in better economic condition. They simply have fewer alternatives and less to lose. Witness the Palestinian-Israeli dispute.

Real economic development will remove some important barriers that otherwise block the achievement of a peacekeeping international economy. A world of developed nations is not inherently a world without war, but it would be a world in which the poor would be less desperate and the rich less frightened that "the great unwashed masses" would abruptly pun them down from their mountaintop of privilege and power.

This fear of the rich motivates much of the forcible, often violent, suppression of the poor. But suppression increases the frustration, anguish, and desperation of the poor, which in turn creates and legitimizes greater fear. Locked in this cycle, there is little chance to prevent permanently the eruption of violent confrontation and avoid a widening war that a world armed with modern weapons of mass destruction simply cannot afford. It is true that the conflict between the rich and poor is at least as much a conflict between the elite and the impoverished people of the LDCs as between those who live in the richer countries and those who live in the poorer countries. But we certainly have seen enough examples in this century alone to understand that struggles that begin within Third World nations can jump beyond their borders and/or escalate into wars in which major military powers become directly or indirectly involved (e.g., the civil wars in Afghanistan, Angola, Ethiopia, Nicaragua, and Vietnam). Real economic development can contribute to world peace and security by helping to break this cycle.

There is no technical reason and certainly no moral excuse for the persistence of widespread absolute poverty in a world as productive as that in which we presently live. The elimination of absolute poverty is in our grasp and in our interest. It is an achievable, security-enhancing goal. Beyond this, sustained Third World development should further reduce the likelihood of war by softening the economic pressures on the wider LDC populations. Also, nations have more to lose by disruption of trade as their economic condition improves. And it is easier to establish mutually beneficial relationships between nations as the gap between them narrows.

In a sense, the ultimate goal of global development is to eliminate the Third World as an identifiable entity. Although international differences in living standards may never be completely eliminated, there is no reason why there must continue indefinitely to be a permanent economic underclass of nations. Given this ultimate goal, some of the institutions discussed above can be seen as transitional rather than permanent. If it does its job well, the Global Bank, for example, should eventually put itself out of business, or at least evolve into a very different institution. The GEF and the GIF should likewise either disappear or mutate into agencies for liaison and coordination of global efforts in these areas, rather than as grant making institutions per se. The value of emphasizing intra-LDC trade should also fade and finally vanish as real development succeeds on a global scale. The success of these extremely useful interim approaches will be measured by how quickly and completely they render themselves obsolete.

International Regulation of Multinational Enterprises

Multinational corporations (MNCs) represent an evolution of business beyond the evolution of political systems. Their ability to coordinate financing, investment, production, and marketing activities worldwide is unparalleled by the reach of any single political authority. To some, they are the most dangerous and powerful exploitative force in the world today. To others, they are the greatest force for encouraging development and international economic relationships that build trust and understanding worldwide.

Multinational business already has created a vastly more integrated and interdependent international economy and has the potential to take this process much further. That is to the good. The problem is that much of this integration has been unduly one sided, resulting from decisions made in pursuit of corporate goals emphasizing the growth of the economic power and size of the corporation as an end in itself. However, there is no necessary reason why MNC decisions could not be geared to the pursuit of the broader and longer term goals that characterize the peacekeeping international economy.

The extent to which MNCs have, on balance, encouraged or discouraged real economic development is still a matter of controversy. Have they contributed additional capital, trained more indigenous personnel, and transferred useful technology to the LDCs? Or have they drained limited LDC capital pre-empted already skilled workers, and transferred only technology that is so sophisticated and inappropriate to LDC conditions that it has little positive spin off to the rest of the economy? Either way, it is important to understand that the effects on development, whether positive or negative, have been incidental to the primary goals of the MNCs. MNCs are not in the business of encouraging or discouraging development they are in the business of making money. However, it is clear that in some cases at least, they have interfered with development directly and/or actively supported repressive governments more interested in law and order than in justice or development. To the extent that MNCs have done this, they have worked in a direction contrary to the purposes of a peacekeeping international economy.

When the narrowly focused interests of private enterprises lead them to take actions that are contrary to the broader public interest, the best way for governments to realign the divergent interests is to restructure the rules of the game under which those enterprises must operate. Ordinarily, this is much cheaper and more effective than trying to intervene in a direct and heavy-handed way in the organization's internal decision-making. For example, suppose a firm were polluting the air with noxious sulfur compounds. The direct interventionist approach might call for a law requiring the firm to burn only low-sulfur fuel oil. But it might actually be much less expensive to burn high-sulfur fuels and use filters to remove the sulfur in the smokestack gases before they leave the factory or possibly to burn a low-sulfur fuel other than oil. Those options and their relative costs are more likely to be known to the firm whose business is at stake than to the government. The restructuring approach, however, might be to institute a fine based on the amount of sulfur that is allowed to escape the stack and to let the firm worry about how to avoid it. This approach changes the rules of the game so that the firm must internally bear the cost of polluting. But it leaves maximum flexibility and freedom of action in the hands of those most able to find an efficient solution to the problem.

To the extent that MNCs are operating at cross-purposes to the principles of a peacekeeping international economy, rules of the game must be established that will cause them to shift to a more desirable behavior without direct intervention. A single world government capable of regulating MNC behavior is not required. It is perfectly possible to achieve a similar result by international cooperation established through mechanisms such as a generally agreed convention on the conduct of multinational business. There are, for instance, a series of such agreements governing the conduct of international air travel, postal service, and behavior on the high seas. To be sure, these agreements are sometimes violated, but on the whole compliance is quite good. Enforcement could be achieved by domestic legal arrangements in each signatory country (the equivalent of multilateral extradition treaties), combined with economic sanctions. Such sanctions are likely to be even more effective against economic enterprises than against renegade governments.

Regulation of International Trade In Hazardous Materials

Because of differences in health and safety regulations among countries, it is not uncommon for MDC manufacturers to sell unsafe or at least questionable products to the LDCs that cannot lawfully be sold domestically for example, pesticides and pharmaceuticals. Such trade, which sends a clear message that people in the LDCs are inferior in the eyes of the MDCs and that they are undeserving of respect for their health and well-being, is not only profoundly immoral but generates the kinds of international antagonism and conflict that a peacekeeping economy must seek to avoid. It is foolish and counterproductive in other ways as well, often exposing MDC consumers of LDC imports to the very dangers their countries have sought to avoid. Therefore, it makes sense to prohibit international trade in such materials; and probably the simplest way is by international agreement, requiring each nation to write into its own legal code a prohibition on the export of any product or material that cannot legally be sold within its own boundaries. Such a law could include the activities of firms that operate outside the nation's political jurisdiction as subsidiaries of firms based within its jurisdiction.

There is a flourishing international trade in another kind of dangerous material as well: finished weapons and the equipment, components, and materials critical to building them. Such trade also is antithetical to a peacekeeping international economy and likewise must be stopped. Trade in materials and equipment critical to the manufacture of nuclear, biological and chemical weapons especially must be restricted, to the extent possible without interfering with non-hazardous civilian use. And except for very small arms, weapons themselves have no meaningful civilian purpose and therefore should be strictly controlled as well. The International Atomic Energy Agency (IAEA), a key element of the Non-Proliferation Treaty of 1968, is set up as an international inspector, monitor, and regulator of civilian nuclear facilities and trade in nuclear materials. Underfunded and understaffed, it should be re-evaluated to see whether it makes more sense to strengthen and expand it or to replace it with a wholly redesigned organization. In any event, whether or not the weapons and weapons material trade encourages conflicts, it certainly increases the likelihood that the conflicts that do arise will explode into violence. It also dramatically increases the level of damage done when war does erupt, as well as the likelihood that suppliers of critical arms to the combatants will be drawn into the conflict. All of this is inconsistent with a peaceful world or a peacekeeping international economy.

International Council on Economic Sanctions and Peacekeeping

Economic sanctions do not generally have a good reputation among specialists in international relations. However, I believe that sanctions can have a useful though minor, role to play in a peacekeeping international economy.

To be effective, economic sanctions must have broad enough support so that they are not easily circumvented. Given the flexibility of trading systems, this is not easy to achieve. For example, when the Carter Administration instituted an embargo on grain sales to the USSR in response to the Soviet invasion of Afghanistan, Moscow simply bought the grain it needed from other sources. If the United States had managed to enlist the cooperation of all the other major grain exporting nations, the Soviet Union would have had a more difficult time. Of course, as long as some other nation that could buy grain from the exporters would have been willing to resell to the Soviets, the embargo would have failed anyway. But if the embargo had been supported by all the major grain exporters, probably the Soviets would have had to pay significantly more for the grain and thereby suffered some economic penalty.

Furthermore, economic sanctions tend to work more slowly than other punitive approaches. Thus the sanctions so far instituted against South Africa have not yet resulted in the abolition of apartheid. However, if these sanctions are as ineffective as some claim, why does the Government of South Africa go to so much trouble to convince people that they are ineffective? If they really are ineffective, would it not be more sensible for South Africa to pretend that they did hurt so that their opponents would be less likely to switch to more effective means of punishment? Economic sanctions create pressure, and that pressure often works slowly. Unless the sanctions are extraordinarily broad and widely supported, they are unlikely to serve as a quick fix. But as an adjunct to positive incentives and other forms of pressure, they still can be useful.

There are three basic categories of economic sanctions: trade embargoes (including the erection of very high, punitive tariffs), financial boycotts (including divestment, disinvestment, and denial of foreign aid), and the freezing of assets abroad. In a global economy embodying the principles discussed in this chapter, it would be exceedingly difficult for any nation to impose an effective trade embargo unilaterally. Critical goods independence and an emphasis on development of multilateral trade provide too many alternatives. Unilateral financial boycotts are also almost certain to have limited effect and for similar reasons with perhaps one odd but notable exception. If one nation has borrowed a great deal from another, the nation in debt may be able to apply considerable pressure to the creditor nation by threatening to cease repayment of the loans. If its debt is large enough, that sort of boycott could pose a real threat to the financial stability of the creditor. Of course, a debtor nation taking such an action might find it extremely difficult to borrow from any potential lender in the future. Unless it has overcome its need for external financing, its ability to carry out a unilateral boycott of this sort successfully is certainly limited. Likewise, a unilateral freezing of assets has a chance of succeeding only if the nation being subjected to the sanctions holds significant assets in the nation imposing the freeze. A nation would be especially vulnerable to such unilateral sanctions if, for example, its investors owned a considerable amount of real estate and other fixed capital assets (e.g., oil refineries, chemical factories, automobile manufacturing plants) located in one other nation. On the whole, multilateral sanctions will be far more effective.

Effective multilateral sanctions require a more or less objective mechanism to trigger the sanctions, as well as an organization to decide which sanctions are most appropriate and to coordinate and monitor their imposition. A Council on Economic Sanctions and Peacekeeping ("the Council"), free of the veto that hampers the United Nations Security Council, could be established within the framework of the UN (with representatives from all the UN member States) or as an independent entity (with the same or similarly inclusive representation), complete with a permanent staff of conflict resolution specialists charged to monitor developing trouble spots, to bring the issues to the attention of the Council, and to make suggestions regarding the formal or informal procedures that might be most helpful in each instance. Any nation that has directly attacked the territory of another nation would be subject to, say, an immediate and total trade embargo imposed by the Council, and the embargo would continue until hostilities ceased and a UN military peacekeeping force could be deployed to monitor compliance with a ceasefire or withdrawal. The sanction would be automatic and mandatory whether or not the offending party is itself a member; and any violator of the sanction imposed by the Council, whether or not it is a member, would likewise become automatically subject to the same sanction until such time as it came into compliance. Further, any member of the Council could petition the body to take action. However, an overwhelming majority (say, three-fourths) probably would be desirable relative to the imposition of sanctions for any domestic or internal behavior (e.g., the practice of apartheid by South Africa within South Africa).

How would it be possible for the Council to enforce its actions? In the same way, I submit, as any international regime is enforced. In the absence of overarching political authority (whether world government or hegemon), compliance with international agreements relies primarily on the value of the agreement to the parties involved. For example, though the 1968 Non-Proliferation Treaty previously mentioned makes no provision for punitive enforcement (indeed, it permits withdrawal at any time simply upon ninety days' notice), no signatory non-nuclear weapons State is known to have violated it. Why? Because of the benefits that are gained as a result of other State parties restricting their behavior in accordance with the treaty. Most international agreements rely to a significant degree on this process of mutual back-scratching. To be sure, the process does not always work. Treaty violations do take place, just as laws in our national societies are broken every day. But in a world more oriented to non-violent international peacekeeping than the one in which we live at present, a world in which progress toward a peacekeeping international economy is well under way, it is not unreasonable to expect that the Council would be seen as an institution with a sufficiently beneficial function to warrant active compliance and support. This might be particularly the case if the Council were authorized also to resort to positive inducements to resolve international conflicts (as did, for example, the United States at Camp David when it found it helpful to provide some financial assistance to Egypt and Israel as part of the conflict resolution process that led to the peace treaty between the two nations). The Council would therefore have the right to petition the Global Environmental Fund, the Global Infrastructure Fund, and the Global Bank to finance a particular project it proposes as a means of settling a potentially dangerous conflict.

Summary

The combination of (1) basically free trade, (2) tariff barriers (if necessary) to achieve independence in minimum critical goods requirements, (3) more LDC trade (including Third World common markets), (4) LDC primary product cartels, and (5) regulation of multinational corporations and trade in hazardous materials may seem an odd mix. It includes elements of free market, protectionist, interventionist, regionalist, globally oriented, and self-reliant development ideology. In a sense, however, this eclecticism is not surprising, as it arises from an attempt to strike two critical balances: a balance between independence and interdependence and a balance in the flow of benefits derived from international trade.

Clear, strong, and concrete emphasis on generating sustained progress in real economic development also is crucial to achieving this dual balance, most particularly over the long run. A Global Infrastructure Fund and a Global Bank could prove to be highly useful transitional institutions in the realization of this goal, while a Global Environment Fund, along with the regulation of international trade in hazardous materials, could help reduce stress on the environment, thereby minimizing ecological conflict. Agreements restricting international traffic in arms, though unlikely to reduce conflict itself, could sharply reduce the outbreak and extent of organized violence. In addition, the Council on Economic Sanctions and Peacekeeping could impose multilateral economic sanctions in such a thoroughgoing and coordinated way as to be far more punishing to violators of generally agreed norms of international behavior than present unilateral or ad hoc multilateral attempts. It also could provide expertise in conflict resolution and broker the use of certain positive inducements to settle conflicts amicably.

We are living in a time of profound change in attitudes, technology, and in the political and economic conditions that surround the institutions of militarism and war. The end of the 1980s has seen remarkable change at sometimes breathtaking pace. In less than five years, relations between the US and the USSR improved enormously, as the Soviet Union underwent a revolution in openness and economic structure more dramatic than anything it had seen since the revolution that brought it into being. After a decade of essentially non-violent struggle, Poland broke through the rigidity and authoritarianism that had held that nation in its grip since World War II. In East Germany, that breakthrough took less than three weeks, and in Czechoslovakia only about ten days. Bulgaria and Hungary also experienced rapid and positive change. Except for the tragic case of Romania, the peoples of Eastern Europe, without raising a gun, without using any of the tools of war, began the process of liberating themselves from the straitjacket of socioeconomic and political oppression under which they had lived for so long. Though it is too early to tell how long-lasting these changes will be, the occurrence of transformations believed virtually impossible only a few weeks or months prior to their actually taking place makes it seem that anything is possible, given the appropriate political will.

4

Program Dimensions and Elements

Introduction

An America-first sentiment in Congress, and initial resistance from the Bush administration, it is remarkable that just five years later Senator Sam Nunn (D-GA) would be able to confidently assert: "History will record the prevention of four nuclear weapons states from emerging out of the wreckage of the Soviet empire as one of the greatest achievements of the decade, " one that laid "an important foundation for a post-Soviet world." More remarkable still, this program was passed in a clear show of bipartisanship after a previous, and somewhat similar, proposal by Representative Les Aspin (D-WI) and Senator Nunn failed just weeks before.

It then conceptualizes Nunn-Lugar assistance, defining it as a defense-related security assistance initiative. Finally, this chapter provides an overview of the logic underpinning each of the program's constituent components and shows how U.S. aid attempts to address the identified needs.

Programmatic Evolution

The ill-fated Aspin-Nunn "anti-chaos" aid initiative set the stage for the later Nunn-Lugar program (see chapter 4). Aspin, chairman of the House Armed Services Committee, proposed on

August 28, 1991, to add a new provision to the pending defense authorization legislation allowing the transfer of up to $1 billion in fiscal year (FY) 1992 Defense Department (DOD) funds for emergency food and other humanitarian assistance to the Soviet Union. Nunn, chairman of the Senate Armed Services Committee, agreed, provided that the money could also be used at the president's discretion for continued Soviet control over its nuclear, chemical, and biological arsenal, as well as for technical assistance relating to defense conversion efforts.

Critics such as Representative Robert Dornan (R-CA) promptly asserted that "a billion dollars is a lot of money," especially for "the former 'evil empire' that caused 46 years of [high] U.S. defense spending." The cosponsors were persuaded to withdraw the initiative even as Aspin steadfastly maintained that it was a "billion-now-to-save-billions-later" proposal, a prudent "insurance policy" that rested on the core assumption that further reductions in the U.S. defense budget would be conditional on the "triumph of democracy" in the disuniting Soviet republics.

While the Aspin-Nunn initiative ultimately failed, the Nunn-Lugar program was successfully established shortly thereafter. Over time, while congressional critics continued in their opposition to cooperative endeavors with Russia and the other new independent states (NIS), the program overcame considerable domestic opposition and initial suspicion in the NIS to achieve, in whole or in part, many of its objectives. For a relatively small sum invested, the Nunn-Lugar program yielded considerable dividends both in the nuclear sphere and in the early post-Cold War effort to secure a democratic peace with the nascent post-Soviet states.

Legislative Guidance

Senator Nunn' s conclusion that "this whole nuclear business was not a zero sum game and that we had better stop treating it as such" was inspired long before by an influential study on the fate of nuclear weapons in an imploding Soviet Union that was presented to him by Ashton Carter and his Harvard colleagues in late 1991. In the wake of the failed hard-line coup attempt against Soviet President Mikhail Gorbachev in August 1991, Nunn saw a real opportunity to help safeguard the security of Soviet nuclear weapons in the midst of profound internal Soviet instability. Under the leadership of Senator Nunn, Senator Dick Lugar (R-IN), and other key congressional advocates, the CTR program began in December 1991. Congress passed the Soviet Nuclear Threat Reduction Act that month, and added to its scope with the Former Soviet Union Demilitarization Act in 1992. These acts provided the initial legislative guidance for the program and allowed for up to $400 million each year in DOD transfer authority to be spent on the purposes specified in the legislation. The combined program authority was codified in the FY 1994 National Defense Authorization Act as the Cooperative Threat Reduction Act, and a new line-item (rather than transfer authority) appropriation of $400 million was authorized. Section 1202 of this act stated that it was in the national security interest of the United States to further five specific objectives. First, the United States should facilitate the transportation, storage, safeguarding, and elimination of nuclear and other weapons in the NIS, including: (a) the safe and secure storage of fissile material derived from the elimination of nuclear weapons; (b) the dismantlement of intercontinental ballistic missiles (ICBMs) and submarine-launched ballistic missiles (SLBMs) and their launchers, and heavy bombers; and (c) the elimination of chemical (CW), biological (BW), and other weapons capabilities. Second, it should assist in preventing the proliferation of weapons of mass destruction (WMD) (as well as their components) and destabilizing conventional weapons in the NIS, and establish verifiable safeguards against their proliferation. Third, it was in the interest of the United States to help prevent diversion of weapons-related NIS scientific expertise to terrorist groups or third countries. Fourth, Washington should support the demilitarization of the NIS defense-related industry and equipment and help enable the conversion of such industry and equipment to civilian purposes. Finally, the U.S. military should expand its program of military-to-military and other defense contacts with the NIS. To carry out the above objectives, section 1203 of this act authorized the following:

1. Programs to facilitate the elimination, and the safe and secure transportation and storage, of nuclear, chemical, and other weapons and their delivery vehicles;
2. Programs to facilitate the safe and secure storage of fissile material derived from the elimination of nuclear weapons;

3. Programs to prevent the proliferation of weapons, weapons components, and weapons-related technology and expertise;
4. Programs to expand military-to-military and defense contacts;
5. Programs to facilitate the demilitarization of defense industries and the conversion of military technologies and capabilities into civilian activities;
6. Programs to assist in the environmental restoration of former Soviet military sites and installations when such restoration is necessary to the demilitarization/defense conversion programs authorized above;
7. Programs to provide housing for former NIS military personnel released from military service in connection with the dismantlement of strategic nuclear weapons, when provision of such housing is necessary for their dismantlement and when no other funds are available for housing;
8. Other programs as described in section 212(b) of the Soviet Nuclear Threat Reduction Act and section 1412 (b) of the Former Soviet Union Demilitarization Act.

To ensure that U.S. funds were not misspent, they were conditioned: no money, goods, or services could be provided until the president certified that the proposed recipient was committed to each of six criteria. First, prospective recipients had to make a "substantial investment" of their own resources for dismantling or destroying WMD. Second, they had to forgo any military modernization program that exceeded "legitimate" defense requirements. Third, the fissile material or other weapons components removed from old weapons could not be used in new weapons. Fourth, recipients had to "facilitate" U.S. verification of any weapons destruction carried out with U.S. assistance. Fifth, states had to comply with "all relevant" arms control agreements. Finally, recipients must observe internationally recognized human rights.

Subsequent legislation also helped shape the program. While the FY 1995 National Defense Authorization Act (P.L. 103-139) largely restated previous program authority, the FY 1996 National Defense Authorization Act (P.L. 104-106) rescinded previous authority that allowed the use of CTR funds for housing purposes. These bills authorized $400 million and almost $300 million in new budgetary authority, respectively. FY 1996 also witnessed the formal transfer of programmatic and budgetary authority for select Nunn-Lugar program components from DOD to the State Department and the Department of Energy (DOE). While DOD's funding was down that year, the combined DOD-DOE-State allocation actually increased to $401 million.

The FY 1997 National Defense Authorization Act (P.L. 104-201) expressly disallowed the use of Nunn-Lugar money for peacekeeping, housing, environmental restoration, job retraining, or defense conversion. However, this bill also contained the Defense Against Weapons of Mass Destruction Act, which reauthorized Nunn-Lugar and again expanded its program authority in other areas. The majority of this "Nunn-Lugar II" money was to be spent for U.S. domestic preparedness for issues related to terrorism and challenges posed by the continuing spread of WMD. As the amendment's chief congressional sponsors observed, this legislation was primarily intended to improve U.S. domestic preparedness for potential nuclear, radiological, chemical, or biological incidents. They also hoped it would enhance the ability of the United States to interdict such weapons and materials at NIS borders and transit points and expand previous U.S. efforts to control those materials at the source. The Clinton administration's entire $327.9 million DOD request for CTR was appropriated, as well as most of the $143 million DOE-State request. An additional $90 million in Nunn-Lugar II funds that the administration did not seek was also appropriated. In all, Congress authorized almost $2.1 billion for Nunn-Lugar between FY 1992 and FY 1996, supplemented by approximately $550 million more for State, DOE, and DOD CTR and Nunn-Lugar II programs for FY 1997. In accordance with long-term defense planning, DOD officials intended to request new funds for each fiscal year at least through the end of 2001. Similar planning by DOE and the State Department called for extending their programs through at least FY 2002 and FY 2003, respectively.

Conceptualizing "Nunn-Lugar"

Advocates such as Senator Nunn, Representative Aspin, leading DOD officials, and other supporters of the program such as Representative Hamilton often argue that CTR "is not foreign aid," not a "gift," but rather an "investment in our own national security."

This is merely creative packaging that helps circumvent the stigma of an often unpopular foreign assistance program. On the other hand, critics such as Representative Gerald Solomon (R-NY) tap into this vein, explicitly referring to CTR outright as the "Nunn-Lugar Foreign Aid Program." Such endeavors, critics claim, are not defense but rather aid programs that should be funded out of the State Department's international affairs account if at all.

In any case, it is evident that Nunn-Lugar bears considerable resemblance to other "security assistance" programs, and while CTR is clearly not a formal component of that long-standing program, it is very much in the nature of security aid and should be characterized accordingly. In each case, resources are largely spent on U.S. goods and services, targeted aid expenditures that endeavor to directly advance the national security interests of the United States.

Clinton administration Assistant to the Secretary of Defense for Atomic Energy Harold Smith attempted to split the difference by defining Nunn-Lugar aid as "cooperative assistance," while at the same time arguing that CTR is "unlike traditional foreign aid" for three reasons: (1) the majority of aid is provided in the form of goods and services; (2) the assistance is issued to help alleviate a clear danger to the United States rather than for long-term political reasons; and (3) the assistance is targeted directly toward specific U.S. objectives and corresponding activities. But in asserting that DOD "provides equipment, training, support, and technical assistance for fulfilling critical objectives," and in recognizing that the vast majority of funds is spent in the United States rather than in the recipient countries, it is clear that Smith could hardly have given a better description of how traditional security assistance programs function.

Defining "Nunn-Lugar"

Initial US-NIS threat-reduction negotiations stemming from the Soviet Nuclear Threat Reduction Act were widely referred to as the Safe and Secure Dismantlement (SSD) talks. However, by the end of 1992, DOD began to refer to it as the Cooperative Threat Reduction program. Despite opposition from the first SSD ambassador, Major General William Burns and other key policy officials in the State Department who considered the DOD term less diplomatically palatable, this term became generally accepted shorthand for Nunn-Lugar. By FY 1994 the term SSD had been phased out and Congress codified the CTR program in legislative form. Nevertheless, it would not be entirely accurate to identify only the DOD programs with the larger collage of cooperative NIS threat-reduction efforts. While DOD maintained that it alone was the program's executive agent, DOE, in particular, undertook a large-scale material protection, control, and accounting (MPC&A) program in Russia and other select former Soviet republics that went well beyond initial DOD-sponsored weapons security and material control programs. Moreover, since DOE's MPC&A programs are funded primarily through its own budget, and since, by law, DOE became the executive agent for these efforts in 1996, it is evident that there are functional and administrative distinctions between the DOD and DOE programs. While the State Department previously had management authority for a variety of nonproliferation programs, particularly export controls and the science and technology centers in Moscow and Kiev, it also adopted budgetary responsibility for them in 1996. Because the scope and composition of U.S. threat-reduction efforts evolved by FY 1996 to include separate programs administered by the Departments of State and Energy, it would be better to expand such a mini-malist definition of Nunn-Lugar-as-CTR into a broader interpretation of the "Nunn-Lugar concept." The FY 1997 National Defense Authorization Act defines "cooperative threat reduction programs" as those that:

- Facilitate the elimination, and the safe and secure transportation and storage, of nuclear, chemical, and other weapons and their delivery vehicles;
- Facilitate the safe and secure storage of fissile materials derived from the elimination of nuclear weapons;
- Prevent the proliferation of weapons, weapons components, and weapons-related expertise;
- Expand military-to-military and defense contacts.

The "Nunn-Lugar program" would be best characterized as this umbrella of cooperative threat reduction programs, a group of joint US-NIS threat-reduction efforts also known as "defense by other means," as Secretary of Defense William Perry often referred to it.

Program Scope and Composition

The Nunn-Lugar program evolved considerably over its life span. While the original legislation envisioned an emergency response to an immediate security need, by 1993 the program became a mainstay of U.S. regional security policy. Due to concerns over political stability in Russia and the expected fallout in strategic matters should hard-liners again come to power or a political crisis escalate out of Soviet President Boris Yeltsin's control, the 103rd Congress (1993-94) effectively institutionalized the program. Subsequently, budgets fluctuated and support variously waxed and waned for each of the three main CTR components: (1) destruction and dismantlement; (2) chain of custody and nonproliferation; and (3) demilitarization and defense conversion. While functionally separate, these areas are "inextricably linked" to one another and, thus, together constitute a larger strategy often called "defense-in-depth."

Destruction and Dismantlement

This is the largest component of the Nunn-Lugar program. CTR funds were provided by DOD to enable or accelerate the destruction and dismantlement of nuclear, and to a lesser extent chemical and biological, WMD in the NIS. Specifically, they provided a key incentive for Belarus, Kazakhstan, and Ukraine to forgo the nuclear option and for Russia to adhere to and comply with the terms of the Strategic Arms Reduction Treaty (START). None of these three states had the financial resources or capabilities to successfully undertake strategic denuclearization. As such, the U.S. provision of mobile cranes, plasma cutters, laptop computers, incinerators to destroy liquid rocket fuel, and similar tools under the Nunn-Lugar program proved materially useful. By 1996, in part because of CTR assistance, approximately 3,800 strategic nuclear warheads were transferred from Ukraine, Belarus, and Kazakhstan to Russia for dismantlement; roughly 1,200 Russian strategic nuclear warheads were dismantled; and 445 ICBM silos were destroyed, complemented by the destruction of 230 SLBM launchers. DOD officials ranked the denuclearization of the non-Russian NIS republics the chief aim and "ultimate yardstick" of the program's success. Indeed, each of these states became nuclear weapons free by the end of 1996.

Russia presented a very different case. Moscow and Washington had common interests in denuclearizing the other three NIS, and cooperated extensively to achieve this objective. At the same time, Russian officials were notably reluctant

TABLE 4.1 Comparative Overview of DOD CTR Funding

Year	Category	Proposed Obligations	Actual Obligations
1994	Destruction/ Dismantlement	475.70	7.67
	Chain of Custody/ Nonproliferation	285.34	96.75
	Demilitarization/ Defense Conversion	150.00	2.01
	Program Support/Other	50.00	11.37
	Total	961.04	117.80
1995	Destruction/ Dismantlement	550.00	241.13
	Chain of Custody/ Nonproliferation	318.04	188.56
	Demilitarization/ Defense Conversion	223.67	120.90
	Program Support/Other	80.27	48.24
	Total	1,171.98	598.86
1996	Destruction/ Dismantlement	706.00	408.24
	Chain of Custody/ Nonproliferation	471.54	363.16
	Demilitarization/ Defense Conversion	167.67	165.73
	Program Support/Other	156.90	112.67
	Total	$1,502.11	1,049.80

Notes: Numerical and percentage calculations are the author's, rounded to the nearest million.

Proposed obligations are those funds for agreements that are still being negotiated, while actual obligations are those that have been reported to Congress, are already negotiated through agreements, and have contracts awarded.

TABLE 4.2 DOD CTR Funding by Country (Current-Year Dollars in Millions as of 8/5/96)

Country	Category	Proposed Obligations	Actual Obligations
Russia	Destruction/ Dismantlement	304.00	181.22
	Chain of Custody/ Nonproliferation	338.26	265.80
	Demilitarization/ Defense Conversion	48.00	47.34
	Program Support/Other	55.55	49.01
	Total	745.81	543.37
Belarus	Destruction/ Dismantlement	33.90	2.51
	Chain of Custody/ Nonproliferation	31.56	23.95
	Demilitarization/ Defense Conversion	25.00	24.70
	Program Support/Other	28.52	20.21
	Total	118.98	71.32
Kazakhstan	Destruction/ Dismantlement	102.00	42.26
	Chain of Custody/ Nonproliferation	46.56	25.24
	Demilitarization/ Defense Conversion	22.00	21.90
	Program Support/Other	1.90	.52
	Total	172.46	89.92
Ukraine	Destruction/ Dismantlement	266.10	182.25
	Chain of Custody/ Nonproliferation	55.16	48.17
	Demilitarization/ Defense Conversion	55.00	54.12
	Program Support/Other	20.03	13.74
	Total	396.29	298.28

Notes: Numerical and percentage calculations are the author's, rounded to the nearest million.

Proposed obligations are those funds for agreements that are still being negotiated, while actual obligations are those that have been reported to Congress, are already negotiated through agreements, and have contracts awarded.

To involve their U.S. counterparts in the actual nuclear dismantlement process. As the General Accounting Office (GAO) reported in 1993, Russia "neither needs nor wants a direct U.S. role in its nuclear weapons dismantlement operations. "Senior U.S. officials were also reluctant, as Major General Burns argued: "I would not want to be responsible for a program where we sent technicians to dismantle their nuclear weapons, even if they asked for it." By the same token, administration officials argued that CTR assistance expedited Russian compliance with the START treaties, a critical objective articulated by DOD officials. Moreover, by 1996 U.S. and Russian officials began to jointly develop Russian technical requirements for CTR aid in the nuclear area. Evident Russian needs and key continuing U.S. interests also raised the prospect of CTR aid, ultimately assisting Russian nuclear arms reduction beyond START I levels.

Russian chemical weapons destruction was a tertiary concern for DOD planners. Indeed, at a cumulative FY 1992-FY 1996 level of $239 million in CTR funding for strategic nuclear arms elimination in Russia alone, the $68 million allocated for CW destruction paled by comparison. This reflected the relative program emphasis to that point. Indeed, the FY 1997 $78.5 million CW destruction request was far greater than the entire amount spent by that time. Just as DOD had higher priorities, Congress proved to be somewhat skeptical of bilateral CW destruction programs. While some 40,000 tons of Russian chemical agent were subject to destruction under the terms of 1989 and 1990 bilateral U.S.-Soviet memoranda and the 1993 Chemical Weapons Convention, this was viewed by key members as more of a Russian than American problem. U.S.-Russian disagreements over the appropriate technology for destruction also slowed the process. By late 1996, however, these problems had been largely resolved, and initial steps toward the design of a pilot facility capable of destroying up to 1,200 tons of chemical agent annually were undertaken. At the same time, however, long-term U.S. resource availability and a mutually acceptable U.S.-Russian division of labor for the final design and construction remained unclear.

On a related front, the FY 1997 administration request included for the first time a small sum for BW facility conversion assistance in Kazakhstan. However, BW destruction would prove more problematic than either nuclear or CW elimination for a number of reasons. Not only is it more difficult to verify BW elimination, but Russia denied repeatedly after 1992 that it maintained an ongoing BW development program in stark contrast to widespread perception in the United States. While limited cooperation in BW facility conversion was eventually undertaken, the scope of such activities would be confined to civilian sites through at least the end of the decade.

Chain of Custody and Nonproliferation

The need to ensure proper control and safeguards over nuclear weapons and weapon components resonates throughout the former Soviet nuclear complex. Chain of custody programs provide assistance to enhance their security during storage and transport to dismantlement and storage sites, at every "link" along the custody "chain." While assembled nuclear weapons are probably less prone to theft or diversion than stocks of fissile material or their constituent components, they nonetheless remain vulnerable at many sites. The most difficult part of building a primitive nuclear device is securing the requisite fissile material; crude bomb designs have been publicly available for many years. As only a modest amount of plutonium or highly enriched uranium (HEU) is required to construct a viable device, the consequences of even a small amount being stolen, sold by a low-paid insider, or otherwise acquired could be devastating. Yet throughout the early 1990s, security and accountability at the more than 100 nuclear facilities in the former Soviet region was, according to one Central Intelligence Agency (CIA) analyst, often no better than a "babushka with a notepad."

In general, administration officials approached the problem of nuclear leakage with a "layered defense" concept. As Assistant Secretary of State Thomas McNamara argued in 1996:

We have sought to control the sources of supply material as well as technology and items used in the production or use of special nuclear materials; to reduce demand; to ease the economic dislocations that followed the break-up of the Soviet Union; and to strengthen law enforcement and intelligence capabilities needed if the security of the material is breached.

Thus, there are many dimensions to the problem of nuclear leakage, and several different Nunn-Lugar programs designed to address these issues. DOD-led chain of custody efforts emphasize weapons security measures and the construction and provision of fissile material storage capacity. Railcar enhancements, the provision of armored blankets, and other similar measures helped ensure the safety and security of weapons in transit. Ukraine, Belarus, and Kazakhstan transferred literally thousands of nuclear warheads to Russia for disassembly per the terms of the START I accord and various other agreements under the aegis of the Commonwealth of Independent States (CIS). One byproduct of the denuclearization process was that Russia claimed it needed assistance in providing adequate and secure storage capacity for the material transferred. While dismantlement was a key U.S. objective, Russian Ministry of Atomic Energy (Minatom) officials pressed early on for greater U.S. attention to increasing fissile material storage capacity. As a result, DOD also worked with the Russian Ministry of Defense (MOD) to provide thousands of fissile material storage containers and to construct a sizable storage facility for warheads awaiting dismantlement.

DOE took the programmatic lead on programs relating to fissile material protection, control, and MPC&A. An ideal nuclear security system would have three essential components. First, physical protection: alarms, sensors, and other barriers designed to deter, delay, and defend against external intruders and insiders removing protected material. Second, material control: locked vaults for nuclear material storage; portal monitors equipped to detect nuclear materials (which prevent workers from carrying nuclear material off-site); continuous monitoring of storage sites with tamper-proof cameras, seals, and alarms; and requirement for personnel to enter sites containing sensitive materials in pairs (a procedure commonly referred to as the "two-man rule"). Finally, material accounting: a regularly updated, measured inventory of nuclear weapons-usable material, based on routine measurements of material arriving, leaving, lost to waste, and remaining within the facility. (This would also presumably include a rigorous control program that would help ensure the accuracy of measurement equipment.) Personnel reliability might also be improved by systematic background checks, training, and reliable salaries for nuclear custodians. External oversight by a regulatory and inspection agency with real enforcement powers would also enhance MPC&A.

Unfortunately, none of these functioned reliably or effectively in the Russian nuclear complex of the early 1990s. Senior CIA official Lawrence Gershwin testified in 1992 that the Soviet system was established for "a strictly regimented closed society worried only about external threats." Accordingly, "security often amount[ed] to little more than barbed wire fences and armed guards, providing scant protection against insiders who hope to get rich by smuggling out nuclear materials. "By 1994, this problem drew substantial attention as DOE began to provide monitoring devices, computers, video cameras, and related equipment in order to bolster NIS (particularly Russian) MPC&A capabilities. While much more work was necessary to develop a highly effective, properly functioning safeguards system, significant progress had been made by the end of 1996.

The State Department complemented the efforts of DOD and DOE. Of particular importance in dealing with the human element in the NIS nonproliferation equation were the science centers in Moscow and Kiev. U.S. officials were concerned by early 1992 that Russian and other officials with weapons-related knowledge would emigrate to such potential proliferant countries as Libya, North Korea, or Iran. In the midst of ongoing social strife and uncertain economic prospects, the centers were intended to help prevent a brain drain of scientific expertise. By the close of 1996 almost 19,000 NIS scientists had taken part in the program, and most external reviews of the centers were very favorable.

Export controls throughout the NIS had not done as well by that time. As Ambassador Michael Newlin observed in 1996, implementation of this component of the Nunn-Lugar program "has been slow." In order to create effective export control systems throughout the NIS, regional officials and U.S. assistance need to facilitate two basic realities. First, they must support the establishment of legal, regulatory, and enforcement systems throughout the NIS. These will help ensure that international transfers of items on the control lists of various international regimes are subjected to governmental reviews and licensing procedures that approve only those exports that are consistent with the regimes in question. Second, U.S. aid must facilitate the creation of decision-making processes that provide adequate attention to the international security implications of exported goods.

Properly functioning export control and MPC&A systems are clearly important in reducing latent NIS nuclear threats. Indeed, given the potential cost of even one incident of substantial nuclear smuggling, the nonproliferation implications could be grave. Given the unstable NIS sociopolitical context and a poorly functioning custodial system, the threat of nuclear leakage became, in effect, the "greatest strategic challenge of our time." While there have been literally hundreds of cases of nuclear smuggling reported since 1992, however, there were only six credibly reported cases worthy of significant proliferation concern through 1996 each involving small quantities of weapons-grade or weapons-usable materials. Nevertheless, as CIA Director John Deutch testified in 1996:

> This threat is real and we should not deny its existence. If a single act of diversion occurs, either the sale of a nuclear device or a meaningful amount of strategic nuclear materials from the Russian complex, we will face a crisis of enormous proportions, and we will devote energy and resources greatly in excess of the cost of a reasonable cooperative threat reduction program would impose on us today. In some sense, making these efforts today is insurance about having to make much larger and much more dangerous resource commitments in the future.

In essence, this nuclear vulnerability of the NIS states is "our vulnerability as well." Concerns over the apparent continuing development of a nuclear black market helped fuel vocal cries for Nunn-Lugar programmatic expansion, particularly in the area of nuclear MPC&A. More extensive international law enforcement cooperation and intelligence sharing to help prevent and counter the prospective illicit diversion of nuclear material was also strongly recommended in a number of widely circulated studies.

Demilitarization and Defense Conversion

This area constitutes perhaps the most controversial dimension of the entire Nunn-Lugar program. The Soviet Union had a large defense industrial complex consisting of 2,000 to 4,000 production enterprises, research and development facilities, and other related institutes that employed between 9 million and 14 million people. In addition, the Soviet Union employed almost 1 million people throughout the large nuclear complex. Together, they posed a critical, if difficult, question: What should be done with the surplus human element and production dimension of the nuclear and defense industrial establishments?

CTR-supported demilitarization programs attempted to address this issue. These efforts undergird the decidedly long-term nature of threat reduction by reducing the capacity and economic pressures in the NIS to continue WMD production, while at the same time constructively engaging the highly specialized individuals who are the key to ultimate success. U.S. defense conversion assistance, its advocates argued, both encouraged the development of a market-based economy in the recipient states and helped prevent the proliferation and sale of advanced weaponry, as well as the incentives for relying on such sales for income.

Defense conversion was not part of the original Nunn-Lugar legislation (although it did comprise one element of the failed 1991 Aspin-Nunn amendment) but was instead added in FY 1993. The following year, Congress established two new defense conversion programs: the Industrial Partnering Program and the Demilitarization Enterprise Fund. To the extent that these efforts provide an economic alternative and, at least in the case of the IPP, are designed to be met by up to 2:1 in matching contributions by private U.S. industry they can reduce pressures for continued defense production for domestic consumption or export and may have the capacity to spin off potentially valuable technology to the U.S. market. As such, advocates of defense conversion investments argue that they are "win-win-win": They help reduce the threats from WMD; they help the NIS build civilian-oriented, commercially viable market economies while reducing excess military capacity; and they facilitate U.S. industry's entry into potentially large markets for civilian goods and services.

In addition to defense conversion, defense and military contacts were added in the FY 1993 legislation. These endeavor to promote transparency through increased information exchange and repeated contact, institutionalize dialogue, and increase mutual understanding between the armed forces of the United States and the NIS republics. For Graham Allison, Assistant Secretary of Defense for Policy and Plans, operationalizing the nascent American-Russian/NIS partnership in security affairs contained at least four related challenges that could be tackled, in part, by a robust program of military contacts: (1) a transformation of the basic posture of Russia and the other NIS toward the West, and the West's toward them; (2) acceptance of partnership in the minds of the leadership in the military and defense communities; (3) construction of a partnership that both engages the personnel and addresses the facilities of nuclear enterprises in Russia and the NIS; and (4) the rapid implementation of nuclear weapons reduction and mutual confidence building.

The extremely low-cost defense and military contacts program was consistently supported by members of Congress, with repeated calls for expanding them both with Russia and the other NIS. Defense conversion presented a very different story. Although such programs had the strong support of senior State, Energy, and Defense officials, private U.S. industry, and the recipient states themselves, many legislators treated these programs as little more than corporate welfare that could be sacrificed on the altar of a balanced budget. While defense conversion programs were generally popular in the former Soviet region (after some initial hesitation), there were severe operational problems with the implementation of such programs. Equally problematic were their political base of support, especially in the United States. Kevin O'Prey rightfully notes that a central defense conversion difficulty was the arduous task of "obtaining adequate funding from a skeptical Congress." While the FY 1993 and FY 1994 legislation established the programs, subsequent legislation de-funded and stripped program authority from efforts such as the DEF. Other programs such as the science centers and the IPP maintained consistently greater, although not unchallenged, support.

Demilitarization efforts such as environmental site restoration or the provision of housing for demobilizing NIS military officers also met with congressional disapproval by 1995. Although these projects had a considerable external advocacy, not least among the recipient states, CTR would not be extensively used for such objectives after 1996.

Other Programs

The domestic preparedness mandate added in the Nunn-Lugar II amendment to the FY 1997 National Defense Authorization Act would fit this category. Concerns over WMD preparedness were bolstered by the World Trade Center bombing in 1993, the 1995 Oklahoma City bombing, and the Aum Shinrikyo cult's usage of sarin gas on a Japanese subway in 1995. U.S. fears in 1995-96 that terrorism was on the rise, that WMD proliferation was reaching to the sub national level and perhaps extending to so-called "rogue states," and that the prospective combination of these two could be detrimental to U.S. national security conspired to foster near-unanimous support for the congressionally sponsored Defense Against Weapons of Mass Destruction Act. As Senator Nunn argued, changes "in the international situation, in access to technology, and in terrorist motivation" mandate serious consideration of the "potential use of unconventional weapons against the United States" or its interests abroad.

Summary

Cooperative threat reduction was a novel approach to the unprecedented problem of maintaining nuclear security concurrent with the collapse of a nuclear superpower. While early efforts focused on the dismantlement, transportation, and storage of tactical and strategic nuclear weapons, later programs evolved considerably beyond this core mission. In part this was because of a fluid, even expansive, NIS threat environment, as initial concerns over the command and control of the Soviet nuclear arsenal gave way to problems relating to the safety and security of the weapons themselves. Over time, other issues, such as the long-term viability of the defence industrial and nuclear establishments, as well as environmental and other demilitarization concerns, found a place under the broad Nunn-Lugar rubric. However, because of their decidedly long-term nature and implicit social-welfare overtones (unlike dismantlement and chain of custody), those issues became politically vulnerable.

There was consensus among most practitioners and close observers that security needs existed that might warrant extensive US-NIS security cooperation. Moreover, many officials and analysts were supportive of some form of a Nunn-Lugar program, even though there was considerable disagreement as to its ultimate duration, appropriate funding level, and specific composition. Indeed, by 1996 no single U.S. government agency implemented "Nunn-Lugar, " per se, but rather its select program components. Moreover, while loosely coordinated, programmatic and budgetary responsibility was spread out among different executive branch agencies in both the United States and the NIS. At the same time, while these various programs shared the common goal of threat reduction, each stressed distinct objectives, such as dismantlement, nonproliferation, and domestic preparedness.

Yet, while there are many differing interpretations as to the scope and breadth of "Nunn-Lugar, " it would appear that two major divergent approaches to the subject emerged. First, there are those who viewed the program in broad or expansionist terms. Nunn-Lugar, argued one Democratic staff member of the House National Security Committee, is an "enormously elastic idea." Another moderate Republican Senate staff member agreed, arguing that Nunn-Lugar was a "concept" or a "tool" for reducing the nuclear threat from Russia and the other NIS. Viewing the program in largely holistic terms, the staffer asserted that he could "live a long time on the sale of an SS-18 and so could four Russian cities."

Others who agreed with this perspective generally argued that a short-term emphasis on destruction and dismantlement, while a laudable objective, would most likely prove shortsighted over time. Also needed, in their view, was attention to restructuring the nuclear and defense establishments, a focus on the fragile social situation in which highly trained specialists were unemployed or underemployed, and intensive long-term attention to sustaining democracy in Russia and integrating it into the international community. This approach had some advocates in the U.S. Congress, in the private sector, and, especially, among key Clinton administration officials; large-scale support for such an interpretation also resided in the NIS.

Alternatively, there are those who viewed the Nunn-Lugar program in more limited, strictly defined terms. Particularly among many congressional critics, the demilitarization and defense conversion aspects of the Nunn-Lugar program were viewed as "jobs programs" or "welfare." For instance, Congressman Benjamin Gilman (R-NY), a self-described "deeply frustrated" chairman of the House International Relations Committee, offered an amendment to the FY 1997 defense authorization bill that would have prevented any U.S. funds from promoting defense conversion objectives in the NIS. Gilman's objections were twofold. First, he was rightly concerned by 1996 that various soft-defense or non defense programs, such as the DEF and the science centers, had been transferred from DOD accounts to the even more cash-strapped international affairs budget. If DOD could not bear the costs for these programs, or was not convinced of their utility, then why should the State Department and its much smaller international affairs account shoulder them particularly in the face of continued downward budgetary pressure? Second, Gilman, like many of his Republican colleagues, remained unpersuaded of the effectiveness or even desirability of some of these programs. This in part resulted from partisan politics, but it also stemmed from a differing calculus of the nature of program needs and how U.S. interests might best be served. The tension between the program's advocates and reformers, on one side, and its critics, on the other, added a layer of complexity to the unfolding political drama of "defense by other means."

5

Cooperative Security: Threat Reduction

Introduction

The wide-ranging problems inherent in Soviet nuclear fragmentation could not be adequately addressed by Russia or any of the other Soviet successor states alone. The proposed U.S. solution to this novel problem set was a legislatively mandated program of cooperative threat reduction inspired by academic work on cooperative security.

The fluid security environment presented by the Soviet collapse posed considerable challenges to policy makers in the United States and elsewhere. The immediate tendency of the Bush administration was to continue, in a rather cautious manner, its policy of supporting the Russian/ Commonwealth of Independent States (CIS) center of what was formerly the Soviet Union. This represented a clear continuation of the status quo ante in which the administration pursued a policy of "tough love" toward the Soviet Union throughout 1989-91. Key elements of this strategy included "locking in" agreements favorable to the United States that would be more difficult to reach with a more hard-line Kremlin regime and conceding nothing that would injure the United States should democratizing reforms fail.

Of central importance to Washington's emerging Eurasian policy was the predominance of Russia in strategic planning, initially at the expense of the other NIS. In practice, this meant the active support of first Soviet President Mikhail Gorbachev and then Boris Yeltsin in U.S. regional policy. Thus, although Undersecretary of State Robert Zoellick testified in late 1991 that the United States needed to actively engage both the Soviet Union and the constituent republics, Paul Goble criticized the administration for pursuing, in effect, a regional policy of "Russia plus branch offices." Similarly, some Russian officials actively referred to the independence of Ukraine and other NIS as a "transitional phenomenon, " warning the American and European governments against opening embassies in Kiev since they would soon be downgraded to consular sections subordinate to their own embassies in Moscow.

The challenge for the United States quickly became one of balancing its relations between Russia and the other NIS. This is particularly evident with respect to Ukraine, a state that, by virtue of its size and potential strength, deserved considerable attention of its own. Indeed, the creation of a modern and independent Ukrainian state was arguably the "most significant geostrategic development" in Europe since the onset of the Cold War. Of particular importance to U.S. policy makers was the near-term disposition and ultimate fate of the Soviet strategic nuclear arms stationed in Ukraine, Kazakhstan and Belarus. The White House stated in clear terms that it would continue to "encourage and support positive, centralized control of nuclear weapons, press for rapid arms control treaty implementation and help reduce [NIS] nuclear arsenals." While a comprehensive treatment of Belarus and Kazakhstan is beyond the scope of this chapter, many of the issues surrounding Ukrainian promises to denuclearize, resistance in the process of becoming nuclear weapons-free, and establishment and broadening of relations with the United States are in many respects largely similar in those states as well.

Denuclearization Imperative

From Ukraine's declaration of independence from the Soviet Union to mid-1993 after Strobe Talbott joined the Clinton administration as ambassador-at-large and special advisor to

Secretary of State Warren Christopher, and became a principal architect of U.S. regional policy Washington approached Ukraine with one key objective: denuclearization. The Bush administration only reluctantly supported Ukrainian independence after it became evident that the Soviet Union had failed as a viable entity and in the face of considerable domestic criticism. It eventually agreed to recognize Ukraine and the other NIS but did not wish to establish deeper relations with them until the nascent republics promised to cooperate fully in the nuclear sphere. Of particular importance, this included the renunciation of displaced Soviet strategic nuclear weapons and adherence to the terms of the Treaty on the Non-proliferation of Nuclear Weapons (NPT).

Early indications over a year before the Soviet collapse were that Ukraine would indeed fulfill its oft-stated denuclearization preferences. As its July 1990 Declaration of State Sovereignty proclaimed, "Ukraine will become a neutral state not participating in any military blocs" and adhere to "the three non-nuclear principles: not to receive, not to produce and not to acquire nuclear weapons." Despite a temporary pause in the transfer of tactical warheads to Russia in early 1992, by May of that year the entire transfer was complete. Moreover, the Lisbon Protocol to the Strategic Arms Reduction Treaty (START) formally pledged Kiev to become nuclear weapons-free by the end of 1994 (see chapter 1). Bush administration officials were initially pleased and offered to spend $175 million in newly authorized Nunn-Lugar funds to assist in the Ukrainian denuclearization process, as soon as Kiev's promise to sign and ratify the NPT was fulfilled.

Ukrainian Foreign Minister Anatoly Zlenko, however, soon lamented that nuclear policy should not be the "epicenter of future U.S.-Ukrainian relations." Indeed, Ukrainian officials recognized early the focused U.S. preference for only one Soviet nuclear successor state and consistently, over time, bid up the price for denuclearization. While Washington extended diplomatic recognition and offered technical and financial assistance for Kiev's nuclear renunciation, that country soon became a "nuclear hedgehog." Ukraine's leaders enjoyed the benefits of a declared policy of denuclearization while at the same time signing but not successfully ratifying START and the NPT until late 1994, and receiving more than twice the initial U.S. offer of Nunn-Lugar aid and considerable other economic assistance in the context of a much broader relationship. It was not until June 1996 that Ukraine finally became nuclear weapons-free.

Ukrainian Concerns and Nuclear Foot-Dragging

A combination of factors conspired by late 1992 (and especially through 1993-94) to call into question Ukraine's repeated commitments to disarm. These included a lengthy state-building process, latent fears of a resurgent and possibly irredentist Russia, a burgeoning popular and official Ukrainian perception that they had been tricked into giving away "their" nuclear weapons, and concerns that non nuclear weapons status would negatively impact Ukraine's international disposition. As Bogdan Horyn, deputy chairman of the newly formed parliament's Military Affairs Committee, stated in 1993: Ukraine "has the right to decide to be a nuclear or a non-nuclear state, " and that previous commitments to become a non nuclear state "were voiced as intentions, not obligations." First Deputy Prime Minister Ihor Yukhnovsky and others increasingly argued that "Ukraine cannot allow its wealth to be taken out of the country free of charge, " a sentiment summarized by Vice President Leonid Kuchma: "What does Ukraine get in return? This is the question that troubles the Ukrainian people and parliament."

While severe Western, and especially U.S., pressure provided key incentives for Ukrainian leaders to denuclearize "in the shortest possible time" under the terms of the Lisbon Protocol, there were also evident counter-incentives to retain the status quo as a de facto or temporary nuclear weapons state. Ukrainian leaders repeatedly voiced concerns over the ownership of the fissile material inside the nuclear weapons and demanded adequate material compensation from Russia and the United States for their removal. Moreover, the threat of a chronically unstable Russia with latent territorial ambitions posed a key security challenge for the newly independent republic.

At the same time, the new Ukrainian government feared becoming "just another Slovakia, " effectively isolated from the West and estranged from the United States. For the leadership in Kiev this would be tantamount to an acute diplomatic snub, arising mainly from national perceptions of Ukraine's rightful status as a significant regional power. Nuclear weapons quickly became a principal vehicle for expressing Ukrainian independent statehood. The Ukrainian government was internally divided, among other things, between those who felt that the weapons might be a useful deterrent and those who concluded that they would be more of a strategic liability. However, there was broad domestic consensus that these weapons could be used as bargaining chips to secure the best terms possible for independence.

Increasingly in relations with the West, especially the United States, policy makers in Kiev asked for denuclearization compensation in four primary areas. First, they requested financial compensation for nuclear removal equivalent to the value of the highly enriched uranium (HEU) embedded in both the strategic and tactical nuclear warheads. Second, they wanted technical assistance and financial backing for the cost of destroying the nuclear missiles. Third, and most difficult to satisfy, Kiev demanded both "positive" and "negative" security guarantees from all other declared nuclear powers since, as various Ukrainian parliamentarians argued, "at stake is the very existence of the state." Positive security guarantees assure that members of the international community will come to Ukraine's aid if attacked, with military force if necessary. Negative security guarantees are, in this context, assurances that external powers will not use nuclear weapons against Ukraine. The leadership in Kiev also extended this principle to the economic sphere with Russia, asking for pledges not to apply "economic pressure" against it for political ends. The intent was clear: Kiev wanted promises from Russia not to attack Ukraine and from the United States that it would not allow such an attack. In response, both the United States and Russia issued limited guarantees to Ukraine. The former committed itself to seeking immediate United Nations action, in accordance with 1968 U.N. Security Council Resolution 255, to any non-nuclear state NPT party that "is a victim of an act of aggression or an object of a threat of aggression in which nuclear weapons are used." Russia, on the other hand, offered a conditional guarantee to "preserve and safeguard the sovereignty of Ukraine and its borders and defend it from nuclear attack."

Not surprisingly, neither of these offers satisfied the Ukrainian leadership. The former was an old guarantee that required NPT accession, while the latter was an offer of protection from Kiev's worst security fear the proverbial equivalent of a school-yard bully offering not to beat it up in exchange for disarming. Kuchma promptly restated the government's request that Washington "bring Ukraine under the protection of its nuclear shield, " even as U.S. officials refused further concessions until START and NPT ratification. Within the year, however, Washington policy makers began to meet Ukraine's fourth condition: a broader relationship with recognition of its prospective future status as a regional power, with less emphasis on nuclear issues, and with the assurance of continued strong bilateral relations in the post-nuclear era.

After considerable strong-armed diplomatic engagement, the three-way atomic diplomacy between Ukraine, Russia, and the United States began to yield dividends. While clearly imperfect and ultimately unacceptable to both the United States and Russia, the Ukrainian parliament ratified the START I Treaty on November 18, 1993, with thirteen conditions. Indicating considerable parliamentary consensus, the lopsided 254-9 vote to pursue "stage-by-stage"' nuclear dismantlement was a compromise position reached between many competing political parties. Specifically, the resolution allowed for the prompt destruction of 42 percent of the nuclear warheads and 36 percent of the delivery vehicles, with the possibility of further warhead and launch vehicle elimination "pursuant to procedures which will be determined by Ukraine, " and then only with "sufficient" external assistance. Parliamentarians also expressed a clear, albeit informal, preference that this should apply to the older, more unstable SS-19 intercontinental ballistic missiles (ICBMs) rather than the more modern SS-24 ICBMs, which could conceivably have commercial space-launch applications. Significantly, this resolution expressly repudiated Article V of the Lisbon Protocol, Kiev's previous pledge to become a non nuclear weapons state "in the shortest possible time." The statement reiterated previous Ukrainian denuclearization demands and asserted ominously that the "prospect of resolution" by parliament on the issue of accession to the NPT would become possible only after START I had been implemented.

Under substantial external pressure, President Kravchuk promptly resubmitted the treaty to parliament, arguing that "if we cannot use or fully control these weapons and cannot replace or service them we must get rid of them." A flurry of diplomatic frenzy between late November and early January set the stage for a more complete agreement between Ukraine, the United States, and Russia. On January 14, 1994, President Kravchuk signed the "Trilateral Statement" along with Presidents Yeltsin and Clinton in Moscow. Under the terms of this executive agreement, within ten months Kiev would transfer 200 nuclear warheads to Russia in exchange for 100 tons of low enriched uranium (LEU) from Russia to be used as fuel for its civilian nuclear reactors; the United States would provide a modest sum of money up front to start the process, to be counted against an outstanding (1993) deal with Russia for the purchase of roughly 500 metric tons of HEU over a twenty-year period. The remaining warheads were to be transferred for an equivalent ratio of LEU within the seven-year dismantlement period originally envisioned for START.

Moreover, Ukraine would receive at least $175 million in CTR funds and other, unspecified U.S. economic assistance, as well as a public reiteration of the previously stated negative security guarantees from Russia and the United States once Kiev ratified the NPT. Additionally, Moscow pledged to "respect the independence and sovereignty and the existing borders of CSCE [Conference on Security and Cooperation in Europe] member states," and provided an explicit reaffirmation "to refrain from threat or use of force against the territorial integrity or political independence of any state." This pledge helped persuade parliament to supersede its previous conditional resolution with an unfettered START ratification in a 260-3 vote. By late 1994 the Ukranian parliament would ratify the NPT.

Thus, in contrast to the official Bush administration position that "we're not going to engage in a bargaining process. We're not going to bid up the price," the Clinton administration did just that, and ultimately achieved its central denuclearization objectives. At the core of Ukraine's strategy between mid-1992 and early 1994 was a deliberate policy of attempting to use the strategic weapons as "bargaining chips" to gain concessions from other states. Indeed,

Ukraine accomplished much through its efforts, particularly in elevating its status and prestige and in securing higher levels of U.S. aid and Russian compensation. But it did not get everything it wanted, especially the credible and strong external security guarantees it requested or U.S.-backed extended deterrence of Russia, before finally translating its oft-stated pledges into denuclearization deeds in early 1994.

Ukrainian state behavior on nuclear issues suggests a latent utility for Nunn-Lugar funds. While they clearly had manifest material importance in the technical aspects of denuclearization, they also became symbolically important to the recipient states as an external confirmation of statehood and a key indicator of their status in Washington. Arguably, NIS officials began to view CTR assistance as a de facto entitlement over time. As the deputy foreign ministers of Belarus, Ukraine, and Kazakhstan argued in 1996, any significant reduction in Nunn-Lugar funds would be considered a "breach of commitment," would "create doubts as to the reliability of the United States as a long-term partner," and might cause "serious delays in the final elimination of nuclear weapons."

U.S.-Russian Strategic Relations

U.S. relations with Russia were of a fundamentally different nature than those with the other NIS. While the question of how, whether, and when to recognize the nascent republics mattered considerably, it was evident that there would be some form of a Russian-oriented Soviet successor state with which Washington would establish diplomatic ties. At the same time, the United States clearly emphasized its relations with the Soviet Union (over Russia, per se) for as long as possible in an effort to sustain the center over the periphery.

Indeed, the Bush administration generally adopted a "Gorbocentric" policy, as Deputy National Security Advisor Robert Gates referred to it. As explained by National Security Advisor Brent Scowcroft, the United States would initially relate to Yeltsin only to the extent that he did "not undermine the center or jeopardize our relations with Gorbachev." The Soviet president, of course, "spared no effort" in capitalizing on the Bush administration sentiment, according to long-time Soviet Ambassador to Washington Anatoly

Dobrinin. Yeltsin endeavored to secure "financial and moral support" for his reforms from the United States and "from Bush personally as a way of supporting his regime and his popularity at home."

While there were vocal detractors on both sides to a close-knit Soviet-American partnership in the face of an emergent "new world order, " the pervasive sense of many was that a strong relationship between the two erstwhile Cold War enemies was not only possible but desirable particularly if the Soviets would continue their seeming progress toward democratization. America's stake in the "second Russian revolution" of late 1991 was such that Harvard scholars Graham Allison and Robert Blackwill proposed, in conjunction with Russian economist Grigory

Yavlinsky, a "grand bargain" of economic assistance for Soviet democratic reforms. In a fairly narrow window of opportunity, the West, and especially the United States, could not afford to "simply watch events unfold" in a changing Soviet Union. Rather, "having spent some $5 trillion to meet the military challenge of the Soviet Union around the globe" during the Cold War, would the United States and its allies "opt out now when the Soviet future is being formed?" Others, such as U.S. Ambassador to Russia Jack Matlock agreed by early 1991 that it was "time to bring our economic might to bear, " but stopped short of advocating a large amount of aid for Gorbachev's Soviet Union.

However, given Washington's paramount interest in the future of the Soviet command and control apparatus and the specter of nuclear fragmentation, assistance in the nuclear sphere had considerable bipartisan support on Capitol Hill and solid administration backing by the time of the Soviet collapse. After an ill-fated attempt by Representative Les Aspin (D-WI) and Senator Sam Nunn (D-GA), chairmen, respectively, of the House and Senate Armed Services Committees, to provide up to $1 billion in "anti-chaos" aid for the Soviet Union, Nunn and Senator Dick Lugar (R-IN) successfully proposed the Soviet Nuclear Threat Reduction Act. Setting aside $400 million in transfer authority from the Defense Department's (DOD) fiscal year (FY) 1992 budget, plus an additional $100 million for humanitarian relief, the Nunn-Lugar program became as instrumental in nuclear engagement with Russia as with the other former Soviet republics.

"Cooperative threat reduction" cut to the core of post-Cold War security relations between the United States and Russia. It was cooperative because of its bilateral character and a burgeoning mutuality of interests underlying certain aspects of the residual Soviet nuclear threat. It was threat-reducing because it was both a means to an end and an end in itself, a vehicle as much for the process of engagement as a practical denuclearization tool. While the provision of materiel proved useful in helping Russia meet its START commitments, equally useful was the framework that Nunn-Lugar established as a springboard for dialogue over the larger state of U.S.-Russian relations.

When faced with the prospect of curtailed Nunn-Lugar funds in 1995, for instance, Russian diplomats cautioned that suspending CTR assistance would cause "serious damage to the broad Russian-American cooperation in strengthening security and the non-proliferation regime." U.S. policy makers were not unaware of the diplomatic significance of CTR programs. As Arms Control and Disarmament Agency (ACDA) Assistant Director Michael Nacht argued, CTR and other Clinton administration nuclear arms control initiatives constituted "an integral part of the American national security policy to draw Russia into both Western and global political, economic, and eventually security institutions."

Strategic Engagement and Partnership

Russia and the United States clearly shared a number of security interests in the disintegrating Soviet Union, particularly with respect to the question of nuclear inheritance. Andrei Kortunov, director of the Center for Geopolitical and Military Forecasts in Moscow, boldly argued in 1995 that the "nuclear interdependence between Washington and Moscow cannot be denied." Similarly, deputy director of the Moscow-based Institute for the Study of USA and Canada, Sergei Rogov, observed that the "need for Russian-American cooperation" derived mainly from five main strategic imperatives: (1) requirements for strengthening strategic stability, including those aspects that directly influence NIS regional stability; (2) the high cost of a continued arms race; (3) the mutual desire to avoid conflated military tensions in geopolitically important areas that could involve Washington and Moscow in regional conflicts; (4) an interest in maintaining an advantage over the military potentials of such possible rivals as Germany and Japan or other new centers of power; and (5) the commitment to strengthening nuclear, biological, and chemical weapons nonproliferation, restricting conventional arms and technology transfers, and controlling the international trade in ballistic and cruise missiles. Of particular importance would be the "joint management of the common nuclear legacy."

From Washington's perspective, it was, ironically, threats associated with post-Soviet Russian weakness that mandated extensive security cooperation. If, as Senator Nunn argued, the strategic calculus was that a fragmented Soviet nuclear custodial complex or command and control apparatus was a net strategic liability for the United States as well as for Russia, then it would be in the security interest of the United States to facilitate and enhance, to the extent possible, secure Russian nuclear custody. Even as the reality of nuclear deterrence continued unabated and the United States adopted a "lead-but-hedge" nuclear posture, the underlying approach became one of "preventive defense" predicated more on U.S. perceptions of changed adversarial intentions than capabilities. Indeed, reductions in capabilities are often viewed as key indicators of altered perceptions of a state's basic intentions, as are such mutual confidence-building measures as the 1994 Russian-American accord to de-target nuclear weapons. The idea of mutual assured survival became a policy preference articulated by key civilian officials in DOD, arguably coexisting with the continued reality of mutual assured destruction.

Building on the legacy of the Intermediate-Range Nuclear Forces and START I Treaties, and in conjunction with a disintegrating Soviet Union, Presidents Bush and Yeltsin met briefly at Camp David in February 1992. Shortly thereafter, they agreed to a "Charter for American-Russian Partnership and Friendship" at the June 1992 summit in Washington, D.C. The charter proclaimed that "mutual trust and respect" would be the basis of their relations and heralded a "determination to promote confidence and enhance understanding" between the Soviet and American peoples. The Soviet and American presidents further agreed that the "well-being, prosperity, and security" of a democratic Russia and the United States were "vitally interrelated." Indeed, the leaders proceeded on the assumption that an expansion of contacts between citizens would "help ensure the irreversibility of the new quality of American-Russian relations, " and secure the democratic peace. Of particular importance, given the "potential for building a strategic partnership, " would be an expansion of defense and military contacts and wide-ranging cooperation in areas ranging from countering terrorism and narcotics trafficking to joint peacekeeping. Another agreement reached at the June summit allowed for the provision of Nunn-Lugar assistance in the transport, safeguarding, and destruction of nuclear, chemical, and other Russian weapons.

In their quest to translate the lofty summit rhetoric into reality, the charter encouraged summit meetings to be held at regular intervals. Subsequent meetings were held bilaterally on an approximate annual basis, and in other multilateral settings (such as G-7 and the United Nations) as well. Some of these summits were noteworthy for the substance of their discussions, even if there was not always tangible progress on difficult issues. At the same time, others, such as the "Nuclear Safety and Security Summit", were designed to address particular issues but instead were more memorable for their high-profile public-relations value for the presidents to help bolster their domestic standing.

To assist the process of dialogue at the highest levels, in addition to presidential summitry, the 1993 Vancouver Summit initiated a process of vice-presidential contacts on roughly a semiannual basis. The so-called "Gore-Chernomyrdin Commission" (GCC) was comprised of eight specialized lower-level GCC working groups dealing with a variety of defense, economic, environmental, energy, and science and technology issues. While not a panacea for bilateral ills, the GCC proved useful in giving some significant disputes a senior-level hearing and in elevating the profile of certain issues on the bilateral agenda. As such, the seven GCC meetings held between mid-1993 and late 1996 provided a forum for constructive dialogue on outstanding and unresolved issues, and in some cases even helped resolve those disputes. However, implementation of some GCC agreements, such as a June 1995 accord between Department of Energy (DOE) Secretary Hazel O'Leary and Victor Mikhailov, head of the Ministry of Atomic Energy (Minatom), to shut down the plutonium-producing reactors at Tomsk-7,

Krasnoyarsk-26, and Chelyabinsk-65, remained problematic years later. Partly as a result of intensive presidential summitry, and in part because of the GCC, U.S. cooperation with Russia on a variety of nuclear and other defense, economic, space, commercial, energy, technological, environmental, and other issues was furthered. By the same token, the United States and Russia, as sovereign states, would of course continue to disagree on key collateral security concerns such as Russian technology transfers to Iran's nuclear and missile development programs or the emergent U.S. preference to modify the 1972 Treaty on the Limitations of Anti-Ballistic Missile Systems.

Nuclear Security Cooperation

Due to its status as the sole Soviet nuclear successor state, and because of the extensive nature of its nuclear complex, the situation with Russia was different from the other NIS. Moscow and Washington had a common interest in realizing the denuclearization of Ukraine, Belarus, and Kazakhstan and worked closely toward that end. By the same token, cooperation between them on Russia-specific nuclear issues achieved a decidedly mixed record through the end of 1996. While significant progress was made in enhancing Russian fissile material protection, control, and accounting (MPC&A), much more work remained. In other areas, such as defense conversion, little of substance was accomplished. The extent and form of U.S.-Russian security cooperation varied widely across issue area and with respect to the actors' institutional preferences, the general state of bilateral relations, and the political acceptability of the act in question.

A diverse cast of bureaucratic actors within Russia participated in Nunn-Lugar efforts. Indeed, as Vladimir Orlov observes, "too often the Russian government appears to speak with multiple, contradictory voices on particular issues relevant to CTR." Alternatively, some Russian officials displayed a remarkable "lack of interest in cooperative agreements that the United States considered equitable and essential." Even when agreements were signed, their effective implementation was sometimes difficult. The Central Intelligence Agency (CIA) repeatedly estimated that the Russian government lacked the physical ability to implement some agreements signed by government officials, resulting to some extent from institutional weakness. Moscow thus bore considerable responsibility for a slow start in some aspects of the Nunn-Lugar program, particularly with regard to projects relating to MPC&A and plutonium disposition.

Ministry of Defense

The MOD weapons complex contained weapons launchers, weapons storage facilities, and an elaborate weapons transportation system. MOD 's 12th Main Directorate, headed by General Evgenii P. Maslin, was responsible for the deployed weapons as well as their safety in transit. Since early Nunn-Lugar program requirements called for the tactical and strategic denuclearization of Belarus, Kazakhstan, and Ukraine, DOD- MOD cooperation was extensive.

Responding to internal Russian critics, particularly in parliament, of Russian-American cooperation in this area, General Maslin argued that the United States offered its assistance "absolutely free of charge. What we can do is accept U.S. aid and thank them for it." While the MOD "would like to carry out the process of nuclear warhead dismantlement by ourselves there is a lack of financial resources in Russia." Neither MOD nor DOD officials pressed for extensive U.S. involvement in dismantling the nuclear weapons themselves, but U.S. weapons security assistance was helpful especially during transportation to and from storage sites. MOD and DOD officials were generally pleased with cooperative efforts in this area, although each side expressed disappointment over their relatively less extensive involvement in meeting the "social problems embedded in demilitarization that must become a top priority."

Ministry of Atomic Energy

Russia's weapons assembly/disassembly, design, and material and component production facilities were all under Minatom 's purview. In all, more than 700,000 individuals were employed in the ten "closed" nuclear cities that consisted of various laboratory or industrial facilities where weapons research or production occurred. This vast labyrinth of research institutes and production facilities designed, manufactured, and maintained Russia's nuclear weapons and also supported naval propulsion and civilian nuclear power programs. It comprised the heart of the Russian nuclear archipelago and, as such, made Minatom a "key Russian agency" in dealing with two central nonproliferation objectives: (1) dismantlement of nuclear warheads and (2) prevention of the proliferation of weapons, their components, or related expertise.

Minatom 's relations with both DOD and DOE were never as cordial as the MOD-DOD relationship. Inside Minatom, the extent and type of cooperation with U.S. agencies varied widely, in part due to "arcane bureaucracy, disorganization, and turf battles." The 4th Main Directorate, which dealt with issues relating to the nuclear fuel cycle (such as the plutonium cities mentioned above), was not generally as successful in cooperative endeavors as the 5th Main Directorate, which was charged with nuclear weapons research and development. The principal tasks that faced the 4th Main Directorate were far-reaching in scope and impacted directly on the very structure of the nuclear complex. Its facilities were involved in agreements to shut down the plutonium-producing reactors, verify plutonium stockpiles, build a fissile material storage facility, and sell HEU to the United States. The 5th Main Directorate, by contrast, had relatively greater success. These projects, such as securing armored blankets, emergency response equipment, and railcar upgrades for transport, were relatively small-scale and involved both modest transfers of equipment and relations more at the technical than diplomatic level. While the 5th Main Directorate appeared to be a well-organized, relatively stable, active participant with its MOD and American counterparts, the 4th Main Directorate was under much greater socioeconomic stress and was more dependent on Minatom 's political leadership and industrial infrastructure. The latter's projects also cut across bureaucratic and interest group lines much more than the former 's, since they related to nuclear issues with commercial and social implications (such as plutonium production).

The post-Soviet challenges facing Minatom were extensive and, presumably, provided a strong imperative for cooperation with the United States on some issues. Indeed, the proposed FY 1995 level of \$64.2 million in U.S. project assistance to the atomic ministry was not only larger than U.S. aid for cooperative projects with the Russian MOD but comprised almost one-third of Minatom 's \$203 million budget. In addition to substantial DOD-Minatom cooperation, by the end of 1996 DOE had established cooperative working relations with eleven sites in Minatom 's civilian nuclear complex and an additional ten in its defense complex. Yet, most officials within Minatom, starting with Minister Mikhailov, appeared to be largely disenchanted with CTR by 1995. In one assessment, Russian officials rated the International Science and Technology Center (ISTC), emergency response equipment and training, enhanced railcar security, and fissile material containers as "good" or "excellent." On the other hand, defense conversion, export control assistance, MPC&A, chemical weapons (CW) destruction, the provision of armored blankets, and cooperation on the fissile material storage facility were "very poor" or just "satisfactory." MOD was more satisfied in general, but the office of the president rated most CTR programs in Russia even lower. This was perhaps due more to unfulfilled expectations of the United States than to a technical assessment of the relative merits or achievements of each program clearly a strong statement on the Russian-American strategic partnership.

Minatom 's early reluctance to engage extensively with U.S. agencies slowed the implementation of various cooperative programs. And after 1993, when cooperation began to improve, a variety of problems conspired to weaken Minatom 's incentive for close ties. Many CTR projects simply became secondary priorities for some of Minatom 's directorates due to the economic crisis in the nuclear industry, particularly the widespread unemployment and social dislocation. Moreover, the overall deterioration of Minatom's human and physical infrastructure combined with an influx of nuclear weapons in accordance with the START I accord to further strain the nuclear complex. Another key problem related to Minatom's leadership: While Mikhailov was initially a cautious supporter of the Nunn-Lugar program, he became increasingly skeptical over time, in part because of clashing priorities for program assistance. In late 1991, for instance, he proposed that the entire $400 million in initial CTR funds be spent on construction of a fissile material storage facility rather than spreading them out among several different projects. This project not only received a small amount of total program funding, but in light of repeated delays, a site change, and an upwardly revised expected operational date of 1998 also took far longer than Minatom officials initially anticipated.

Cooperation between Minatom and U.S. agencies varied considerably on other issues. For example, in the case of nuclear smuggling, the U.S. government had to first persuade the Minatom leadership of the severity of the problem. Through early 1994, Mikhailov and others denied that a problem existed and continued to resist significant external collaboration for even longer, in part because it directly challenged their institutional competence. Alternatively, Mikhailov explained away Western "accusations" of nuclear leakage as "somebody's fantasy or a special forgery. There are many in the West who want to tarnish the nuclear industry in Russia's nuclear complex." At the same time, the U.S. National Academy of Sciences declared the problem of nuclear leakage "urgent, " and thousands of reported cases of thefts of nuclear material were reported internationally many allegedly coming from Russian nuclear installations. Some of Minatom's other activities, such as the 1995 sale of nuclear reactors to Iran and continued civilian nuclear cooperation with Cuba, antagonized American officials even as they brought much-needed hard currency to Russia. On other issues, particularly the brain-drain problem, Minatom 's perspective was arguably more in line with U.S. policy.

To deal with these issues, two distinct but complementary approaches developed to U.S.-Russian nuclear security cooperation: top-down, government-to-government diplomatic efforts and bottom-up, laboratory-to-laboratory technical cooperation. The former provided an overarching framework for U.S.-Russian relations. The latter proceeded more informally with the blessing of their parent agencies and was based largely on cooperation between the technical staff of DOE's national laboratories and Minatom's 5th Main Directorate.

While necessary, government-level cooperation often proceeded slowly. With respect to the establishment of the ISTC in Moscow, for instance, the United States, Japan, and Germany announced with Russia in 1992 their respective intent to establish Western-financed science centers in Moscow and, later, in Kiev. The objective of the centers was to help alleviate the potential for an emigration of scientific personnel with advanced and specialized weapons-related knowledge to such potential proliferant countries as Libya, Syria, North Korea, or Iran. Negotiating agreements for the establishment of the centers took months, as they became bogged down on such issues as the proposal review process, tax exemption for wages paid to NIS scientists, the diplomatic status of center personnel, and intellectual property ownership. However, once the agreement was signed with Russia, its implementation was again delayed during the agreement ratification processes in the Russian legislature. The Supreme Soviet refused to even take action on the ISTC for a year after the agreement was signed. In October 1993, acute political differences between Yeltsin and the communist-dominated legislature led to his dismissal of the parliament just as it was preparing to consider the ISTC agreement. One month later, he signed a presidential decree granting provisional approval of the agreement. Within just five months of Yeltsin's action the ISTC Governing Board held its first meeting in Moscow and approved $15 million in scientific projects.

To help offset the delays inherent in formalized government-to-government contacts, because of the time-sensitive nature of the tasks involved, and to circumvent the bureaucratic resistance of certain ministries and departments in both Russia and the United States, DOE's national laboratories instituted a deliberate bottom-up cooperation policy to complement the top-down efforts in the MPC&A domain. The directors of the Los Alamos and Lawrence Livermore national laboratories extended invitations to their Arzamas-16 and Chelyabinsk-70 counterparts to visit in February 1992, and reciprocal site visits occurred the next month. This began the process of establishing quasi-independent, less-formal relations that, while still subject to national-level policies and formal governmental acquiescence, grew considerably over the next several years. Indeed, DOE officials were so pleased with the so-called lab-to-lab approach that what began in 1993 as a $2 million experiment became a $15 million per year program just two years later. This constituted more than a sevenfold increase in funding commensurate with the deepening laboratory-to-laboratory relations.

Other Cooperative Efforts

Outside MOD and Minatom, weapons-usable HEU was used as fuel for some naval and civilian merchant ships and in space, research, and other reactors; and plutonium was produced by and recycled into a modest number of breeder reactors and used for certain research purposes. These also posed significant proliferation risks. While cooperative efforts were of a decidedly lower near-term priority than in the larger weapons complex, by the end of 1996 DOE had established cooperative relations at seven independent civilian nuclear sites and three naval nuclear fuel sites. Bureaucratic responsibility for these programs, as well as those relating to defense conversion and others, was widespread. Russian agencies such as the State Committee for Defense Industries, Federal Service for Hard Currency and Export Control, Ministry of Finance, Ministry of Foreign Affairs, and others each played a role in negotiating or implementing the complex CTR equation. More significantly, GAN was formally given the authority as early as June 1992 to regulate nuclear safety in all Minatom and MOD nuclear facilities. However, it maintained a decidedly low profile in the face of considerable bureaucratic strength elsewhere.

Politics of Distrust: Partnership Reconsidered

The U.S.-Russian "strategic partnership, " established formally at the June 1992 Washington summit, deteriorated considerably over time. While Secretary of State James Baker, and later President Clinton, hailed the U.S.-Russian "strategic" partnership, by 1993-94 Russian Foreign Minister Andrei Kozyrev referred to it as the "lagging" partnership and questioned whether Washington and Moscow could move beyond a "cold peace." Secretary of Defense William Perry referred increasingly after 1994 to the "pragmatic" partnership, and Secretary of State Warren Christopher suggested with his Russian counterpart in 1995 that the partnership was "maturing." Christopher's successor, former U.S. Ambassador to the United Nations Madeleine Albright and Kozyrev 's successor, former intelligence chief Yevgeny Primakov, did not manage to bring the partnership any closer. There developed over time a notable tendency for Russian and American officials to agree when to disagree and to accept some policy divergence as the normal science of diplomacy. Others referred to the "limited, " "premature, " or "virtual" partnership between the United States and Russia. Similarly, Robert Blackwill recanted his earlier advocacy of the "grand bargain, " arguing in 1993 that "we should put aside talk of partnership between Russia and the United States. The objective conditions are no longer ripe for such an ambitious understanding, if they ever were."

Part of the reason for this growing estrangement between Moscow and Washington was the painstakingly slow pace of progress on tangible issues of perceptible common interest. For instance, while the United States pressed for greater transparency in and verification of Russian nuclear dismantlement, DOE was initially reluctant to accept Russian requests for reciprocal openness with respect to its own nuclear facilities. Washington initially took the approach that the problem of arms-control compliance and the threat of nuclear leakage was on the Russian, not American, side; therefore, Moscow should make what amounted to unilateral concessions. Moreover, the slow pace, modest amount, and fluctuating levels of Nunn-Lugar and other aid disappointed Russian government officials. And a continuing downward spiral in domestic Russian socioeconomic conditions combined with distress over unrealized visions of massive Western financial assistance to spoil previous expectations of close ties in other issue areas. The science centers, for instance, only became fully operational in 1995, well after Secretary Baker's 1992 pledge to expedite their establishment. This was a pattern reflected across the board in terms of aid disbursement.

Another reason for U.S.-Russian divergence was each side's evidently differing strategic calculus. Indeed, U.S.-Russian discord on collateral security issues profoundly influenced the overall state of bilateral relations, challenged the extent to which each side's security interests were indeed mutual, and dampened prospects for longer-term cooperative security relations. On the issue of expanding the North Atlantic Treaty Organization (NATO), for instance, U.S. officials repeatedly stressed to Russian officials "that it is not in Russia's interest to take an antagonistic approach to the issue enlargement need not be a zero-sum game for Russia." If, however, Russian officials define their interests differently, such argumentation is politically problematic: Russian, not American, officials determine Russia's national interests and policies. Indeed, Russian officials feared that the West was taking unilateral advantage of Russia in its moment of weakness, as Russian Defense Minister Igor Rodionov articulated in 1997: "There is a very real risk" that as NATO continues to expand "every possible effort will be made to deprive Russia of its strategic nuclear weapons."

There were many other areas of disagreement, such as the 1995 Russian sale of nuclear reactors to Iran and Cuba, mutual recriminations over high-profile espionage scandals and the 1996 U.S. decision to continue developing and eventually deploy highly advanced theater and possibly national missile defense systems. These and other issues conspired to diminish Russian interest in cooperating with Washington, even as a Republican Congress became increasingly skeptical of cooperative endeavors with Russia. U.S. strategic relations with Russia became so strained by 1996 that some close observers felt they were almost "out of control."

An expansive near-term threat environment mandated continued substantial cooperation on such matters of mutual interest as the continued safety and security of Russian fissile material. But there were manifest limits to such close cooperation, as evidenced by the gradual degradation of the U.S.-Russian strategic partnership in the face of both differing external security views and the genesis and development of domestic political reward structures in both Russia and the United States that increasingly seemed to favor hard-line decision makers. As the so-called "partnership" waned, the moderate and liberal elements of each society were pressured by those in Russia who viewed it as a "great power": derzhavniki, or, in Western terms, statists. Similarly, in the United States those, especially in Congress, who viewed Moscow with inherent suspicion or favored unilateral U.S. defensive strength over the prospect of cooperative security endeavors rose in stature over time.

While Moscow and Washington realized certain gains through their partnership, the prospects for U.S.-Russian security cooperation were ultimately constrained both by differing international interests and by legislative and bureaucratic politics on each side. Whatever remained of the grand partnership by 1996, it was evidently not a relationship between equals. As Sergei Rogov expressed it, the "West is acting as a winner dictating new rules of the game." Even as Russia pursued a foreign policy less deferential to the United States and the West by 1994, U.S. policy was increasingly fused with the practical considerations of power asymmetry and strategic dissonance. While voicing the lofty rhetoric of cooperative security, the Clinton administration pursued a decidedly different course. Despite repeated proclamations by Soviet and then Russian officials that there were "no winners and no losers" at Cold War's end, history may ultimately judge their claims otherwise.

Summary

The "political and economic chaos" that followed in the wake of the Soviet Union's demise "mitigates against such far-reaching cooperation, " at least in the short term, and "vastly complicates" planning for any viable cooperative security regime over the long term. Clearly, this conundrum will trouble "those in the West who long for a simple solution to this complex problem."

Applied to the priority U.S. objective of denuclearizing the non-Russian former Soviet republics, Washington pursued a strategy laden with "carrots" and "sticks." While the Bush administration emphasized the latter, U.S. officials eventually relied more on positive incentives to achieve their denuclearization objectives. Diplomatic recognition and support, trade and economic incentives, and substantial technical assistance were provided along with Nunn-Lugar aid to facilitate the desired denuclearization outcome, even as the program helped establish an overarching framework for U.S. security engagement with the NIS. While Belarus, Kazakhstan, and Ukraine were to differing degrees hesitant to denuclearize, and Russia was reluctant to engage in extensive nuclear cooperation with the United States, the Nunn-Lugar program helped pave the way.

It is axiomatic that "the greater the perceived vulnerability, the stronger the incentive to seriously explore a cooperative arrangement, " particularly under conditions of pervasive mutual vulnerability. Indeed, as long as the United States feared a fragmented, decentralized, or unstable Russian nuclear command and control system, or perceived a security risk in the threat of "loose nukes" in the NIS, it had a clear incentive to cooperate with its former adversaries. But the expanding threshold of potential common threats merged with collateral areas of security disagreement within the context of the overall relationship. Together, these eventually made it unclear for how long and to what degree Moscow and Washington would jointly endeavor to tackle even the common problems related to nuclear security.

A U.S.-Russian "Grand Bargain" ultimately proved beyond reach, while the prospects for a long-term program of robust U.S.-Russian nuclear security cooperation became increasingly uncertain through the end of 1996. Still, while it would be "difficult to attach precise costs" for realizing what might be termed the "Grander Bargain" of NIS denuclearization, some scholars estimated in early 1993 that it would likely total "several billion dollars annually." In practical terms, this would constitute just a "few percent of the hundreds of billions of dollars" that the advanced industrial democracies would spend for defense over the next decade. When the final score is tallied, however, perhaps the grandest bargain of all was the successful denuclearization of three NIS and a modest enhancement of Russian nuclear material protection over the first five post-Soviet years all for a modest two-tenths of one percent of annual U.S. defense expenditures.

6

Policy Process in Nuclear Security

Introduction

This was the rough equivalent of "bringing together the people who bought you the $1,000 coffee pot with those who brought you the gulag." While laden with sarcasm, such a metaphor underscores the intensive executive-level politics underlying Nunn-Lugar 's programmatic directions, priorities, and outcomes.

The new independent states (NIS) of the former Soviet Union (FSU) bear considerable responsibility for the initial lethargy that characterized program implementation. As Mitchell Reiss observes with regard to the Ukrainian case, the combination of "erratic leadership, personal rivalries, and unclear policy produced a series of uncoordinated, halting steps toward a declared but distant goal of nuclear disarmament." Similar statements can also be made about the Belarussian and Kazakhstani efforts to denuclearize, as Belarus did not even have a Ministry of Defense with which to work until at least mid-1992, and Kazakhstan did not even sign the requisite agreement enabling work with the United States until December 1993. The unstable domestic political environment in each of these states presented a variety of challenges: sweeping personnel changes, the lack of qualified personnel assigned to nuclear tasks, the establishment and ongoing consolidation of national security bureaucracies, frequent or severe executive-legislative altercations, and other tumultuous exercises. Collectively, these essentially guaranteed that progress would be slow at best.

Of all the NIS, the Russian case is the most complex, largely because of the greater magnitude of the challenges faced and the commensurate requirement for more extensive bilateral Russian-American security cooperation. For many U.S. participants and observers, it is virtually axiomatic that Nunn-Lugar activities initially progressed slowly, due in large part to hesitant or reluctant Russian participation. Indeed, one prominent study of U.S.-Russian nuclear cooperation asserts in no uncertain terms that the initial delays of more than two years were "due primarily to a lack of interest in Moscow." However, while substantial attention has focused on the former Soviet states, comparatively little energy has been spent examining domestic politics on the U.S. side.

This chapter seeks to redress this shortcoming by examining the U.S. executive branch process, with particular attention to the major players in the "policy game": the White House and National Security Council (NSC), the DOD, the Department of Energy (DOE), and the State Department. While the Arms Control and Disarmament Agency (ACDA), the Department of Commerce (DOC), the intelligence community, and other federal agencies also played supporting roles, space considerations prevent detailed treatment of their specific functions or missions here.

Executive Branch Politics and CTR

The Nunn-Lugar program was deeply affected by a continuously shifting interagency process. Although DOD was designated by statute as the executive agent for the CTR program, DOE became the executive agent for the material protection, control, and accounting (MPC&A) element of the Nunn-Lugar program in fiscal year (FY) 1996. Similarly, the State Department acquired executive authority over the science centers and shared responsibility for NIS export controls with the Commerce Department. The efforts of these departments were coordinated primarily via an interagency working group (IWG) led by the NSC, but there were other IWGs that focused on such specific issues as MPC&A, export controls, nuclear smuggling, plutonium disposition, and other issues relating to select portions of the Nunn-Lugar program.

The NSC spearheaded, in conjunction with the ambassador for safe and secure dismantlement (SSD), Nunn-Lugar efforts in the initial policy-setting or diplomacy stage of the program, which began in late 1991 and was gradually deemphasized by the beginning of 1994. Subsequently, the interagency process for coordination of overall U.S. government assistance to the NIS over time fell increasingly under the purview of the special advisor to the president on assistance to the NIS. Day-to-day responsibilities for implementation remained, nonetheless, with DOD and the other responsible executive agencies. Moreover, the ultimate lines of authority were further muddled as the FY 1997 National Defense Authorization Act mandated the creation of a White House special coordinator for nonproliferation and significantly expanded Nunn-Lugar 's scope to include domestic preparedness for possible contingencies involving weapons of mass destruction (WMD). The evolution over time of relations between the NSC and the Departments of Defense, Energy, and State, as well as continuing changes in the interagency coordination process, bore directly on both the achievements and shortcomings of the Nunn-Lugar program.

Indeed, "bureaucratic politics," or the "process by which people inside government bargain with one another on complex public policy questions," shaped the Nunn-Lugar program at practically every phase of its development. Disagreements over funding and programmatic priorities within and among the various executive branch players help explain the program's perceived slow start. By the same token, what developed into a clash of institutional preferences largely account for the "balkanization" of the program in FY 1996, a compromise in which responsibility for several Nunn-Lugar program elements was transferred from DOD to the Departments of State and Energy.

Yet, those responsible for the genesis, policy development, and implementation of the Nunn-Lugar program were not mere garden-variety bureaucrats who pursued "turf, " or an imperialistic expansion of their professional domain. Rather, many of these individuals were agenda-driven members of different "epistemic communities," or networks of specialists "with recognized expertise and competence in a particular domain and an authoritative claim to policy-relevant knowledge" within that issue area. While emphasizing complementary (and, to some extent, sequential) goals, the Nunn-Lugar program was deeply affected by policy entrepreneurs who would pursue dismantlement and denuclearization, on the one hand, and nonproliferation objectives on the other. The clash between these ideas more specifically, the specific hierarchy of objectives and exact nature of the tasks at hand in an effectively zero-sum budget climate led to clashing institutional preferences as members of these disparate groups became officials in different agencies or departments under President Clinton.

Even before the Clinton administration came to office, the Cooperative Denuclearization Working Group at Harvard University's Center for Science and International Affairs (CSIA) had developed an ambitious policy agenda for NIS denuclearization. Some members of this group, in conjunction with colleagues at the Brookings Institution, Carnegie Endowment for International Peace, and Stanford University, were also generally supportive of a "cooperative security" framework for post-Cold War U.S. foreign policy. Several of those who participated in one or more of these projects became officials in the first Clinton administration and some of them acquired direct responsibility for Nunn-Lugar efforts: Robert Bell and Coit Blacker joined Dan Poneman on the NSC staff; Graham Allison, Ashton Carter, Michele Flournoy, Laura Holgate, Catherine Kelleher, and William Perry went to DOD; others, such as Jane Wales, went to other executive agencies; and some, such as Ken Myers, continued in a congressional capacity. Particularly in the Office of the Secretary of Defense (OSD), these officials brought to bear a well-defined policy agenda, and, during their time in office, endeavored to translate their largely predetermined preferences into policy. They placed heavy emphasis on, among other things, the safety and security of the Soviet nuclear arsenal, NIS denuclearization, U.S.-Russian military and defense relations, and defense conversion.

Other academic, scientific, and policy professionals collectively formed a more loose-knit epistemic community that focused on a variety of nonproliferation issues. Prominent individuals associated with the Federation of American Scientists, the Union of Concerned

Scientists, the Natural Resources Defense Council, and elsewhere (primarily in or affiliated with Washington, D.C.'s nongovernmental policy community) suggested a different set of nonproliferation priorities and offered alternative policy recommendations for the incoming Clinton administration. The policy agenda proposed by the nonproliferation group was in some respects similar to the one emphasized above: How should the United States facilitate the denuclearization of Ukraine, Belarus, and Kazakhstan and secure their compliance with the terms of the Treaty on the Non-Proliferation of Nuclear Weapons (NPT)? However, in other respects, this loose collection of analysts focused on a range of different US-NIS arms control and nonproliferation issues: verifying the elimination of nuclear warheads, cutting off the production of fissile material, disposing of highly enriched uranium (HEU) and plutonium, sharply reducing the U.S. and Russian nuclear arsenals, facilitating the Russian chemical weapons dismantlement process, enhancing NIS export controls, and reducing the flow of conventional arms from the NIS.

Just as this group was arguably less cohesive than the dismantlement community, their agenda was also very broadly defined across a wide range of issues. Nevertheless, they directly affected administration policy and the US-NIS arms control agenda. Influential specialists from this community, such as John Holdren and Frank von Hippel, became deeply involved in President Clinton's Committee of Advisors on Science and Technology (PCAST). Others, including Jessica Stern, Matt Bunn, and Ken Luongo, became officials in the White House Office of Science and Technology Policy (OSTP), NSC, DOE, and elsewhere. Moreover, the views of some, such as Thomas Neff and Thomas Cochran (among others), who did not officially enter office, were repeatedly considered at the highest levels. On the Russian side, senior officials such as Victor Mikhailov and Valeri Bogdan of the Ministry of Atomic Power and Industry (later Minatom), Sergei Kortunov and Alexei Manzhosov of the Soviet (later, Russian) Ministry of Foreign Affairs, and others also took an active interest in a variety of NIS nuclear issues and interacted regularly with the U.S. nonproliferation community. While the dismantlement group became involved primarily with DOD, the nonproliferation community strongly influenced the Clinton administration in DOE and OSTP (and, to a lesser extent, NSC and State), as well as their Russian colleagues in Minatom.

There were, of course, other Clinton administration officials and influential outsiders who were not affiliated directly with either of these groups, yet contributed directly to policy in each of these areas. But from the start of the Clinton term, DOD pursued a somewhat different policy course than, at times, other administration officials in DOE, OSTP, NSC, and State might have preferred. In the Nunn-Lugar case, the "pulling and hauling that is politics, " was complemented by an organizational process that was reinvented over time in response to both differing institutional preferences and the sequential logic of an evolving NIS threat environment.

Balkanization became the solution to these differing policy preferences. Indeed, despite the term's overtly pejorative connotation, it eventually became a successful strategy for encouraging DOE and State to "step up to the budget requirements inherent in maintaining [congressional] support for expensive and challenging nonproliferation programs."

Assessing the Interagency Process

The Nunn-Lugar program changed considerably from President Bush's last year in office through President Clinton's second term. A perpetually evolving NIS political environment made the former Soviet region a moving target, while the lack of domestic consensus over regional policy made coherent long-term planning difficult at best. At the same time, the various NIS nuclear and security-related challenges mandated different types of negotiated cooperation among distinct institutional actors. In the end, the Bush administration managed to forge a fragile consensus among the principal U.S. actors that enabled the program to begin, however slowly. This consensus eventually collapsed in the first Clinton term, as, paradoxically, program implementation began in earnest. In neither administration was the interagency Nunn-Lugar process without significant internal problems.

Bush Administration

Some critics assert that while the Bush administration developed a generally effective arms control process mechanism, it did not function well in the case of Nunn-Lugar. To some extent, this resulted from the unprecedented nature of the tasks at hand, so neither the policy nor the organizational process was in place as NIS nuclear challenges arose. It also resulted from the legislative mandate that authorized DOD to transfer funds from other, already programmed, defense accounts to the newly established, congressionally mandated Nunn-Lugar program. Not surprisingly, this triggered substantial political resistance from DOD and other administration officials, which prevented the program from being fully launched until the Clinton team came into office. Bush administration NSC Director for Defense Policy and Arms Control Heather Wilson even claims that in a "classic example" of interagency politics, "no senior official in the Bush Administration actively supported Nunn-Lugar, but every agency wanted to be in charge of it."

This is clearly an overstatement. While Secretary of Defense Richard Cheney was certainly skeptical of the program at the outset, as was President Bush, Secretary of State James Baker was personally involved and quickly interested in the program's objectives and prospective diplomatic utility. However, the White House ultimately accepted the congressional initiative only after being promised that it would retain discretionary expenditure over the funds.

Wilson's indictment of Washington politics focused particular attention on DOD and the interagency process. According to her, the Pentagon "ceded control" of Nunn-Lugar to an IWG, a move that "spelled disaster for rapid action," since it was "entirely incapable of the rapid decision making necessary for a program like Nunn-Lugar." This presupposes an a priori understanding of the program as an emergency initiative, and there is, indeed, evidence that many in Congress viewed the effort this way. The decidedly long-term nature of the Nunn-Lugar program, rooted as it was in a fluid threat environment, evidently did not fit Wilson's image of how the program should have developed. Similarly, she argues that "by bogging the program down in bureaucracy," senior administration officials could "ensure the program's failure" while at the same time making "some progress that was politically sustainable with Congress." The vast majority of those interviewed for this project differ with Wilson, however. Indeed, virtually all interviewees agreed that the process, while both imperfect and in a continuous state of flux, "worked." Moreover, contrary to her assertions, some of those most intimately involved in the program sought diligently to facilitate the program's success.

Clinton Administration

Of course, the Clinton administration was not immune to similar criticism from insiders. Frank von Hippel, Assistant Director for National Security at OSTP (1993-94), concludes after his time in government that the Clinton administration was characterized by an "arms-control-indifferent configuration." Von Hippel asserts that the NSC Defense Policy and Arms Control Directorate "tended to side with the Pentagon on arms control issues." Moreover, intra-administration debate was further stifled, in his view, by the State Department's tendency to follow DOD's lead, a point which led him to conclude in 1995 that there was "very little effective resistance" in the administration to ideas that "undercut" arms control.

Von Hippel's assertion that "there is little opportunity inside government to achieve fundamental changes in policy" undoubtedly stems, in part, from his own admittedly turf-devoid position. Nevertheless, it also substantially reflects the nature of the executive branch political process. First, testimony, speeches, and cables require interagency and White House clearance before delivery, a cumbersome and time-consuming process. In addition, issues not resolved at lower levels are often eventually addressed at higher levels, generally by individuals with less expertise (and often with the assistance of "talking points" prepared by lower-echelon staffers who themselves did not agree). Finally, frequent turf battles, reorganizations, and personnel changes constrain the prospects for continuously effective interagency collaboration.

The logical conclusion in von Hippel's appraisal is that the president, who in theory can redirect policy, is often "forced to accept and become the spokesperson for the lowest-common-denominator consensus that the interagency process produces." With some exceptions, he laments that the process was "very cumbersome and demoralizingly difficult to get anything done." Dealing with NIS nonproliferation challenges required the cooperation of three large, and not always cooperative, bureaucracies: DOD, whose accountants "set almost impossible requirements" for the release of Nunn-Lugar funds; DOE, which was responsible for carrying out elements of the program; and Minatom, a "Soviet-era ministry trying to survive in the post-Soviet market-driven world."

Von Hippel's reflections on the early Clinton administration arms control process and initial CTR program performance are substantially shared by others. This raises an intriguing question: Did Nunn-Lugar ultimately achieve its modestly successful track record because of the interagency process or in spite of it?

Interagency Process: Principal Institutional Players

The NSC has been organized and utilized differently by successive presidents. Because the NSC was created with intentionally vague statutory language, presidents have broad flexibility to tailor it to their particular needs. While the NSC system performs (or should perform) several functions, its central task is to bring together the principal departmental (and other) voices on policy issues, help establish coherent national as opposed to strictly institutional priorities, and integrate individual programs into the broader context of U.S. foreign policy.

In performing these functions, history suggests that a well-managed and effective decision-making process is vital to the task of proper NSC coordination. Without a clearly defined and well-orchestrated process, lines of communication within the NSC and White House hierarchies and between government agencies may fail; implementation of policy directives will likely be problematic; and effective high-level oversight of policy actors can become virtually impossible.

The interagency Nunn-Lugar (or SSD) working group was chaired by an NSC staff member in the appropriate regional directorate and made up of the principal agencies involved in Nunn-Lugar activities: DOD (including representatives from OSD and the Joint Staff [J-5, Plans and Policy]), DOE, State, ACDA, and the intelligence community. This group was charged with coordinating the various agencies involved in NIS threat reduction efforts. But it shared some of the policy responsibility with other working groups often under the leadership of NSC staffers in the defense policy and arms control directorate dealing with various other specific issues, including fissile material control, nuclear smuggling, and chemical weapons (CW) destruction. Regional policy was also formulated in this IWG and, to a lesser degree, in the Policy Steering Group chaired by Deputy Secretary of State Strobe Talbott. Other working groups devoted to specific issues, such as plutonium disposition or nuclear smuggling, were created as necessary and performed with varying degrees of success.

The most contentious work done by the Nunn-Lugar IWG was the allocation of CTR funds. Each agency brought to the table its own notional budget, often quite different in focus and emphasis. Not only did the IWG determine precise funding amounts for each component of the program (e.g., denuclearization, MPC&A, demilitarization) but also the levels of funding for each of the recipient states. At times, the IWG's determinations were further altered in high-level White House negotiations. Moreover, sometimes specific program guidance was established and coordinated by the IWG, only later to be modified through negotiations with NIS governments. And often Congress helped clarify or determine program objectives by statute or informal feedback to administration officials. Although the NSC chaired the SSD IWG, DOD largely dominated the group since it held primary budgetary and implementation responsibility.

In explaining the interagency decision to balkanize Nunn-Lugar, Rose Gottemoeller, NSC Director for Russia, Ukraine, and Eurasia from January 1993 to December 1994, recalls that the primary initial goal of the Nunn- Lugar program was the denuclearization of Ukraine, Belarus, and Kazakhstan. She persuasively argues that it was achieved with "spectacular success" in large part because the White House "exercised full flexibility to use" the program. This notwithstanding, she concludes that such efforts today would "be more likely to fail, for Nunn-Lugar and other related assistance have dispersed into the government departments responsible for implementing the program."

Gottemoeller also asserts that the early Clinton administration was "not fully aware" of two fundamental realities: (1) the "consensus supporting the program inside the professional bureaucracy was fragile" and (2) working with the Russians would be time-consuming and, in the end, would "further erode the consensus in the Washington agencies." According to Gottemoeller, early interagency agreement focused on the smaller-scale, more prosaic projects meant to address the near-term difficulties associated with the transportation or storage of NIS nuclear weapons. Many of these near-term needs, such as the provision of armored blankets, could be readily met with already-existing, off-the-shelf government equipment. The larger, big-ticket items and high-visibility projects (such as the fissile material storage facility) that were intended to serve as symbols of the U.S. commitment to and partial sponsorship of NIS denuclearization, had only modest collective support. The fragile, yet workable, consensus brokered by Gottemoeller and others eventually unraveled, but many of its specific objectives were ultimately realized.

At the same time, the Clinton NSC was criticized for its dual management of nonproliferation and arms control policies. Under Bush the arms control and defense policy directorate had the lead on many, but not all, Nunn-Lugar issues. Under Clinton, however, the directorate charged with managing NIS regional policy spearheaded CTR efforts on the premise that more regional expertise was necessary to resolve what were essentially arms control or nonproliferation issues. While some NSC officials observed that the "issue-sharing arrangement" between the nonproliferation and regional staff worked well, others, such as Natural Resources Defense Council analyst Thomas Cochran, argued that there was poor coordination between the two. While perhaps organizationally questionable, it nevertheless appears that the generally good personal working relations among NSC officials in the two directorates often smoothed over flowchart pitfalls.

While not all IWGs proved equally effective during the Clinton administration and the nature of NSC "leadership" varied considerably with individual staffers, those NSC offices with jurisdiction over Nunn-Lugar program elements generally earned high marks, contributing positively to the NSC's role as coordinator and facilitator. Some IWGs were effective in maintaining program cohesiveness in light of the sometimes widely divergent programmatic goals of DOD, State, and Energy. With strong ties to the Office of the Vice President and the Office of Management and Budget, some NSC officials were able to bring high-level weight to bear when necessary.

Over time, the Nunn-Lugar IWG diminished in importance as program implementation began in earnest. By 1996 this IWG met only once a year instead of monthly, as in 1993. To a great extent, this was because many of the outstanding policy issues had been largely resolved, although there were evidently residual interagency policy squabbles over such large-scale projects as the fissile material storage facility and CW destruction that would continue well beyond 1996. After the policy-setting phase largely came to a close, and the program later fragmented, lines of authority and oversight became increasingly muddled. Implementation responsibility often devolved to the level of the individual executive agencies; and other offices, such as State's special coordinator for the NIS, assumed an increasingly prominent role.

Department of Defense

The Soviet Nuclear Threat Reduction Act made DOD the executive agent of the Nunn-Lugar program, on the assumption that the department would rapidly implement the program in pursuit of time-urgent objectives. Yet the Bush and Clinton administrations treated the program quite differently and, consequently, it performed dissimilarly. In effect, DOD demonstrated through Nunn-Lugar that it can either effectively delay or rapidly implement a program or policy.

Bush Administration

Senior Bush administration officials initially saw the Nunn-Lugar program as a premature reduction in defense spending. Since the Cold War could be rekindled under a neo-authoritarian, revanchist regime in Russia, administration officials argued that it would be better to wait and see how the evolving political situation of a disintegrating superpower developed before committing to significant reductions. The White House also complained that the program was congressionally imposed and, as such, constituted legislative micromanagement of foreign policy. Moreover, it objected that the program contravened its "tough love" policy toward the evolving Soviet Union. Nowhere else in the administration were these views as deeply held as in DOD under Secretary Cheney's leadership.

To be sure, reluctance to provide assistance to recent enemies of the United States, the growing domestic calls for a "peace dividend" to be reinvested in non defense endeavors, and the administration's general tenor of "prudence" inspired only modest early support for the Nunn- Lugar initiative. Perhaps most important, however, were initial budgetary constraints. In the program's first two years, authorizing legislation only permitted transfer authority. In practice, this involved a programming trade-off: every dollar spent on the CTR program meant one less for a program or policy priority already identified by the Bush administration. It is not surprising, then, that Secretary Cheney reportedly called it a foolish program, or that the Pentagon was quickly criticized for a perceived pattern of "foot-dragging" during its first two years.

While NSC officials saw Nunn-Lugar in the context of a broad policy initiative of which CTR was a part, DOD officials viewed the situation quite differently. Indeed, initial CTR outlays were reprogrammed from already-budgeted items in the Operations and Maintenance account and took the form of off-the-shelf deliverable goods (often excess DOD equipment) rather than newly purchased items. This was the fragile consensus that emerged during the Cheney era. As one member of the Joint Staff argued, Nunn-Lugar programs "meant a less capable U.S. military, so J-5 wanted it to physically produce a less capable FSU nuclear threat. That tradeoff we could live with." Strategic nuclear weapons dismantlement thus became the order of the day for Pentagon military strategists, even as civilian officials in OSD were concerned by the interagency discussion over and division of DOD's resources.

Major General William Burns (USA, Ret.) had the diplomatic lead on SSD negotiations for the U.S. government during the first fifteen months of the program. Together with a cadre of lower-ranking DOD officials, he saw great promise in the program, and did what he could despite Cheney's opposition. During Burns' tenure, many of the Nunn-Lugar efforts were devoted to securing "umbrella" agreements with the constituent republics, and the general focus of the program was on defining specific assistance projects. While DOD lawyers insisted on such comprehensive agreements before issue-specific work could begin in earnest, DOD contracting personnel relied on the standard, complex, and cumbersome contracting procedures that virtually guaranteed little progress could be made in a timely fashion. The lengthy contract-awarding process, particularly within the defense establishment, essentially ensured that only a small amount of money could be spent within the first couple of years even if the recipient states and U.S. negotiators had been able to agree quickly on funding priorities. Some DOD officials dispute that contracting procedures could have been significantly expedited due to congressional concerns that the money be spent in the United States; the concerns of U.S. companies that they be allowed to fairly compete; diplomatic delays in negotiating specific contracts; and, especially, resistance from the DOD Comptroller's Office to spending the money at all. Others simply suggested that "we use the mechanisms at our disposal to spend the funds," without inventing "unnecessary" new procedures.

Indeed, Bush administration DOD and State Department officials evidently agreed that measuring the success of CTR by the obligation rate of funds was a poor yardstick. Stephen Hadley, Assistant Secretary of Defense for International Security Policy, testified before Congress that it was DOD's plan to "identify early on an increment that makes sense, that may be modest in dollars, but gets the program underway." Similarly, Hadley's counterpart in State, Assistant Secretary Reginald Bartholomew, argued that "we are very conscious of the fact that we need to be able to demonstrate that it's a sensible expenditure of [taxpayers'] money" rather than spending it just "for the sake of spending it."

While there was clearly some DOD resistance to the CTR program in both the Joint Staff and OSD during the Bush administration, it is equally true that there was some base of support that allowed the program to move modestly ahead. While important, the "foot-dragging" charge should not be overstated for two reasons. First, negotiating in the context of the unstable and fluid governing structures of the emergent NIS was time-consuming and difficult. Yet the initial efforts of key Bush administration defense personnel laid the foundation for later successes. Second, although the initial Nunn-Lugar legislation specified DOD as the executive agent for the program, its officials apparently intended to support State's diplomatic lead. As Hadley testified just a few months after the program began, DOD intended to implement the "process that

Ambassador Bartholomew has been heading and coordinating within the government." DOD would "help define what needed to be done and what kinds of support and assistance we should provide, and then we would take over and go ahead and deliver it." DOD thus intended from the outset to be the chief implementing agent of the program. At the same time, destruction and dismantlement issues, particularly the safe and secure transportation and storage of the weapons in question, became the project's early focus in no small part because of DOD policy preferences.

Clinton Administration

When Les Aspin became Secretary of Defense under newly elected President Clinton, OSD was substantially reorganized. Throughout most of his brief tenure as Secretary of Defense, Aspin 's top-level management remained caught up in the sometimes grueling process of Senate confirmation. But many of the personnel he brought on board, as well as a substantial portion of his organizational redesign, remained intact under his successor: Deputy Secretary of Defense William Perry. Indeed, the CTR program developed an increasing sense of organizational coherence under Secretary Perry. An evident chain of command between the secretary and the two principal offices with organizational responsibility for the policy and implementation (or acquisition) sides of Nunn-Lugar emerged. On the policy side, Assistant Secretary of Defense for International Security Policy (ISP) Ashton Carter became the "Nunn-Lugar czar." Under him was the Special Coordinator for Cooperative Threat Reduction (SC/CTR), first Gloria Duffy and later Laura Holgate. In addition, within ISP two other offices had an interest in CTR: the Deputy Assistant Secretary of Defense (DASD) for Russia, Ukraine, and Eurasia Elizabeth Sherwood, and the DASD for Threat Reduction Policy, Susan Koch.

Still, turf fights on the policy side of DOD, together with a somewhat questionable organizational structure with fluid ISP management responsibilities, affected DOD's overall CTR performance. To some extent, this resulted from the bifurcation of management and policy oversight responsibilities. To illustrate, the division of labor between the special coordinator's and Koch's offices was somewhat unclear. Duffy was brought on board specifically in her capacity as the "Special Coordinator" for CTR, a deputy assistant secretary-level position intended to emphasize the seriousness with which the Clinton team viewed CTR, with the underlying organizational vision that she would represent DOD in negotiation efforts led by SSD Ambassador James Goodby at State. Titled "Threat Reduction Policy, " Koch's offices handled a much wider portfolio than just Nunn-Lugar activities. But it is unclear exactly why Koch was directly responsible instead of the special coordinator for an office titled "Cooperative Threat Reduction." In practice, it appears that the SC/CTR had de facto jurisdiction over that CTR office even though it existed administratively under Koch's division, since most of the Nunn-Lugar work done there had to be cleared through the SC/CTR's office anyway. Organizationally, however, both Koch and Duffy reported directly to Ashton Carter.

Koch's formal job description only vaguely described her position and responsibilities with respect to Nunn-Lugar. The DASD for threat reduction policy "formulates and implements policy measures whose primary purpose is to advance the safety, security and dismantlement of former Soviet nuclear weapons, " and "works with the Special Coordinator in formulating and implementing policy regarding the Nunn-Lugar program." The SC/CTR, on the other hand, had a more finely tailored (if not completely clear) description of duties relating to the program. She was responsible for the "formulation, oversight, and implementation of selected policy areas" related to Nunn-Lugar, and would provide "study support and advice as directed." Moreover, the SC/CTR would represent DOD as deputy head of the SSD delegation and was charged with coordinating both budgetary matters and industrial outreach; she would also be CTR 's congressional liaison. While day-to-day management responsibility for individual program elements dismantlement and chain of custody, demilitarization and defense conversion, and military and defense contacts was splintered among the three offices in question, it also appears that overall Nunn-Lugar management, leadership, and direction was ensconced in the SC/CTR 's office.

While the NIS deserve major responsibility for the program's slow start, another factor was the internal structure of DOD's program management. Given the simultaneous need in 1993-94 to focus on

NIS diplomacy, coordinate policy with other executive branch agencies, and maintain support on Capitol Hill, it is arguable that the first SC/CTR 's mandate was simply too large, staff support too modest, or both.

As the NIS diplomacy phase of the program waned in 1994, the second SC/CTR, Laura Holgate, was able to devote more attention than her predecessor to marshaling a supportive constituency (and mollifying critics) on Capitol Hill and to addressing the concerns of program critics inside and outside of government. At the same time, Major General Roland LaJoie (USA, Ret.), a senior defense official responsible for implementation, drove that effort forward and further enabled the SC/CTR to engage in interdepartmental and congressional diplomacy. Moreover, the ISP chain of command became clearer as Holgate reported directly to Koch rather than to the assistant secretary, as before.

On the acquisition side, Assistant to the Secretary of Defense for Atomic Energy Harold P. Smith, Jr. had oversight responsibility for CTR implementation. Carter and Smith worked closely together in coordinating the program, undoubtedly contributing to the program's substantial advancement. In order to further expedite the rate of fund obligation, the CTR Program Office was established in 1994. Led by LaJoie, who reported directly to Smith, this office centralized the acquisition process, freeing the Defense Nuclear Agency (later, the Defense Special Weapons Agency and then the Defense Threat Reduction Agency) to deal with the nuts-and-bolts issues of securing and managing contracts. The On-Site Inspection Agency, which also reported to Smith, was responsible for CTR audits and examinations.

The congressional committees with jurisdiction over (or interest in) CTR or NIS regional policy had trouble understanding the executive branch organization for CTR and became frustrated by the perceived lack of DOD attentiveness to such concerns. Moreover, one of the earliest and most persistent complaints against the Nunn-Lugar program was the initial slow expenditure rate of funds. Congress became so exasperated with the "glacial pace" of program implementation that it refused to renew almost $330 million in expired FY 1992 and FY 1993 transfer authority. After the Republican victory in the 1994 mid-term election, the program came under increased scrutiny by legislators who were variously opposed to cooperative endeavors with Russia, concerned about the adverse impact of CTR on the defense budget, or convinced that the program had been ineffective. After program implementation gained considerable momentum in 1994-95, some of these concerns were alleviated even as new challenges were posed by an influential group of skeptical legislators.

In March 1994 congressional testimony, Smith explained the program's slow start as a function mainly of the inevitable time lags associated with international negotiations, arguing that mutual trust between the United States and its former adversaries took time much more than just two years to develop against a backdrop of fifty years of suspicion. After his tenure as SSD ambassador, Major General Burns reflected that "we had to rid ourselves to some degree of Cold War mindsets and so did they. Then we had to pick up from basically blank slate and devise a relationship so we could talk comfortably about nuclear weapons, something we could not talk comfortably about before." The Russian Federation signed its umbrella agreement in June 1992, Belarus in October 1992, Ukraine in October 1993, and Kazakhstan not until December 1993.

The slow pace of negotiations with the former Soviet republics certainly offers a prime explanation for the program's lethargic beginning. In the program's first year, for instance, $1.7 million of the initial $400 million in Nunn-Lugar program funding was spent on executive branch travel to the NIS, while most of the remainder was not spent. These negotiations were both time-intensive and necessary. However, critics in Congress and in the NIS argued that this constituted little more than "nuclear tourism" in which the "main local beneficiaries of American aid were luxury hotels billing an endless stream of visiting delegations at the rate of $300 per night."

In response to such congressional and recipient-state criticism, Carter declared that 1993 was an "extraordinary year" for the CTR program, with twenty-five bilateral implementing agreements signed in the FSU by March 1994. According to Carter, the CTR program relied on a "one-two punch" to push forward the U.S. objective of denuclearizing the NIS: first diplomacy, then implementation. If 1993 was the "year of negotiations, " then 1994 would become the "year of implementation." There was a nine fold increase in obligations between 1994 and 1996 alone, in no small part due to the efforts of LaJoie's CTR program office. Thus, although program implementation could be (and was) significantly expedited, program formulation (including NIS diplomacy) in reality began shortly after the passage of the Soviet Nuclear Threat Reduction Act; expectations of more rapid progress were arguably unrealistic or misplaced.

Overall, the sustained, high-level support given by both Aspin and Perry (the latter's "favorite program"), as well as substantial commitment by the principal officials involved, proved essential for the effective implementation of the CTR program. Even though few of the key senior civilian officials in OSD with responsibility for CTR had lengthy management or bureaucratic backgrounds, and consequently hit a steep learning curve in many of the mundane minutiae of bureaucracy such as budgetary submissions and reporting requirements they quickly became more effective than the previous administration. Although the charge of 'foot-dragging' was applied to the Cheney Pentagon, Perry's DOD not only effectively administered the program but also demonstrated a significant learning capacity. After Congress refused to renew the expired FY 1992-93 transfer authority, Under Secretary of Defense for Policy Walter Slocombe established a legislative working group comprised of "each component in Policy with a legislative affairs interest."

Balkanization Grand Bargain

CTR was initially financed exclusively with DOD funds and its implementation was managed by DOD officials. As such, DOD became the proverbial "900-pound gorilla" of the interagency process. This invariably meant considerable Pentagon influence on Nunn-Lugar programmatic priorities, primarily the dismantlement of strategic nuclear weapons but also defense conversion and, to a lesser extent, demilitarization. Even before FY 1996, however, both DOE and State informally led individual DOD-funded Nunn-Lugar programs dedicated to objectives complementing the primary goal of denuclearization. While they supported the continued reduction in strategic nuclear launch vehicles, both pressed hard for less attention over time to strategic nuclear weapons dismantlement vis-a- vis an increased emphasis on alternative objectives. State, in conjunction with DOE, and with solid support from OSTP and key

NSC staff, lobbied to emphasize nonproliferation activities. Their attempts to secure a significant amount of CTR program funds for the design and development of a fissile material storage facility and MPC&A did not sit well with DOD officials, who generally emphasized different policy goals in the context of scarce resources.

Fissile material control and accounting ranked a distant third in DOD's hierarchy of Nunn-Lugar program objectives, behind the denuclearization of Ukraine, Belarus, and Kazakhstan and assistance for Russia in reducing its strategic nuclear arms to Strategic Arms Reduction Treaty (START) I levels by 2001. Even when NSC personnel stepped in to help establish CTR funding priorities, an act that happened repeatedly despite DOD cries of micromanagement, the Pentagon often carried the day. DOD officials did not necessarily disagree with increased attention to MPC&A but argued that other agencies should manage and fund it. Indeed, they recount tremendous congressional pressure to not reprogram allocated funds or "give" them to other agencies. This added a clear incentive for senior DOD officials to support programmatic fragmentation with the underlying logic that agencies should request funds directly from their own congressional authorization and appropriations committees for the specific projects or programs they wished to pursue.

At the same time, State, DOE, ACDA, Commerce, and other executive agencies demonstrated a clear preference to rely on a declining defense budget for their various policy goals. Whereas these agencies "could describe in programmatic terms" what they would like to see done in warhead dismantlement and material control issues, "they had no ready funds available. The answer to the budget problem, whatever the project or agency initiating it, tended to be 'Nunn-Lugar.'"

The balkanization grand bargain was part of Presidential Decision Directive-41, "U.S. Policy on Improving Nuclear Material Security in Russia and the Other Newly Independent States," issued in September 1995. Brokered informally in late 1994 by Deputy Defense Secretary John Deutch, this compromise transferred nine different CTR program elements to other federal agencies, effective in FY 1996. Illustrating the deep internal divisions among administration officials, one White House official derided this as

DOD's preferred "let's take our toys and go home" strategy, while another argued quite the opposite: that "having DOE and State spend DOD's money was always a recipe for interagency disaster."

Policy officials throughout the executive bureaucracy agree that the fragmentation of Nunn-Lugar was done largely at the insistence of the DOD. While the official rationale for the programmatic fragmentation was to "ensure that these efforts are more effective and efficient, " DOD officials said privately that the transfers permitted DOD to "focus its efforts on core defense programs and, thus, ensure their most competent and efficient implementation". Indeed, Ashton Carter recalls that when the White House attempted to give technical and programmatic guidance, the effect was to hinder the implementation of Nunn-Lugar. You cannot implement programs, especially the unprecedented programs of Nunn-Lugar, from the White House. You have to implement them from the agencies that have the technical and acquisition expertise, and on-the-ground presence, to make things happen. This was not a case of the White House coaxing a reluctant agency to do the right thing: we in DOD were fully committed to the program. We in DOD frankly regarded some on the White House staff as one of the barriers we had to overcome to implement the Nunn-Lugar vision. White House micromanagement was one of the factors that made Nunn-Lugar take off slowly.

Balkanization ultimately helped appease internal DOD critics in the Comptroller's office who endeavored to scale back the Nunn-Lugar program to minimize its perceived adverse impact on the defense budget. It also helped alleviate the worries of OSD officials who, although committed to a modest defense conversion program, were concerned that excessive attention to such projects would foster an undesirable backlash on Capitol Hill. Finally, it also pacified the Joint Staff, whose members fought against the notion of DOD's budget as a "cash cow, " or bottomless balance sheet with which to fund the projects for, or meet the priorities of, other agencies to the detriment of DOD's more traditional war fighting mission.

Department of Energy

In terms of program management, DOE initially played a very limited or supporting role but larger than DOC or ACDA in the

Nunn-Lugar program. Under the Bush administration leadership of Admiral James Watkins (USN, Ret.), DOE initially adopted a slow-going, "wait-and-see" approach to the Cold War's end. Indeed, DOE was overshadowed by the more powerful DOD and State Department bureaucracies, which, in the view of Ken Luongo, were "wedded to the comfortable, but confrontational Cold War method of conducting business with Russia." Moreover, the end of the Cold War prompted widespread calls for a reassessment of the department's role in an emergent post-Soviet world. Internal and external reconsideration of DOE's proper roles and missions as well as those of the ten national laboratories under its jurisdiction became widespread, particularly after Hazel O'Leary became Energy Secretary in 1993. Within DOE, the Office of Arms Control and Nonproliferation (NN-40) took the organizational lead on the Nunn-Lugar program, representing departmental interests at the various interagency meetings and formulating DOE's policies on Nunn-Lugar issues.

Although it occupied a "very passive" seat at the interagency bargaining table at least through early 1994, DOE's technical expertise was crucial to Nunn-Lugar efforts. Initially, DOE's contribution to the program consisted largely of loaning technical personnel for such program elements as the International Science and Technology Center (ISTC) in Moscow, and providing coordination between the various federal agencies in Washington and the involved national laboratories. Given their considerable weapons expertise essential to select CTR efforts, laboratory personnel worked directly on Nunn-Lugar projects such as developing storage containers for fissile material and recommending such emergency response equipment as armored blankets to protect shipments of nuclear weapons components for disassembly.

At times, coordinating these efforts resulted in institutional conflict. The national laboratories are government-owned but contractor-operated. As part of that relationship, the DOE laboratories agreed to perform work handed down from DOE 's headquarters and regional offices. However, under what are known as "work for others" arrangements, the laboratories may engage in contractual work with other public and private entities. Fulfilling contractual agreements under work-for-others arrangements sometimes created management tensions between DOE's Office of Arms Control and Nonproliferation and DOD's ISP office. The former viewed the national laboratories as DOE assets and, accordingly, argued that DOE should have a strong voice in laboratory oversight. The latter, on the other hand, believed that since DOD funded the national laboratory work, the Pentagon should be able to interact directly with the laboratories without DOE "interference" and have complete oversight regarding laboratory work on Nunn-Lugar projects.

DOE's national laboratories also became tentatively involved with their Russian counterparts late in the Bush administration. In May 1991, Under Secretary of Energy John Tuck observed that while the 1954 Atomic Energy Act had been rewritten (P.L. 101-510) to allow meaningful cross-national contact between those who work with the U.S. nuclear arsenal and those who work for foreign governments, the president had not yet made a determination authorizing it. The directors of the Lawrence Livermore and Los Alamos National Laboratories had issued invitations to their opposite numbers in Arzamas-16 and Chelyabinsk-70 to visit the U.S. facilities. This was reciprocated the following month. While a modest beginning, this set the stage for DOE's later extensive involvement with Russian agencies at both the laboratory and governmental levels.

On other issues, DOE was either self-deterred or hampered in its nonproliferation efforts by the interagency and arms-control negotiation processes. In the case of reciprocal U.S.-Russian nuclear site inspections, for instance, DOD and State Department heavy-handedness meant that little progress would be made on such cutting-edge nonproliferation issues as transparency and verification. In late 1993, some U.S. officials throughout the executive bureaucracy, not just in DOE resisted addressing the highly contentious issues of reciprocity. At the same time, Russian officials became more willing to allow U.S. officials access to such facilities as the plutonium-producing reactors at Mayak but only if reciprocal access was granted to U.S. nuclear facilities as well. Deriding this "automatic requirement for reciprocity" as "old think, " Assistant to the Secretary of Defense for Atomic Energy Robert Barker testified in 1992 that a "concern about Russian nuclear weapons security should not result in a mandate for Russian inspection of U.S. facilities." By mid-1994, however, Russian officials visited the plutonium facility in Hanford, Washington, and DOE officials visited Mayak.

Institutional Preferences and Policy Evolution

Some DOE officials argue that their views were often marginalized on the NSC-led, but DOD-dominated, Nunn-Lugar interagency working group. But by mid-1994, on Secretary O'Leary's orders, DOE began to play a much larger role in policy formulation and in its direct implementation. While DOD's generally cautious attitude toward the multi agency expenditure of "its" CTR funds presented an early obstacle for DOE to overcome through 1994, DOE and State also began to diverge on policy issues, or the tactics employed to pursue them, by 1993-94.

One recurring example of Nunn-Lugar interagency bargaining is the case of material control and accounting (MC&A) and physical protection (PP) of fissile material. From the outset, DOD wanted Nunn-Lugar program emphasis on nuclear weapons security and dismantlement. It soon became clear, however, that the safety and security of the dismantled warheads, and especially of the centralized control of the nuclear fissile material, was a major problem for the more than 100 centers in Russia and the other NIS responsible for storing the material. Indeed, some analysts observed that "potatoes were guarded better" than several nuclear facilities in Russia, particularly those under Minatom 's or civilian jurisdiction. Moreover, even the successful reduction in strategic nuclear arms fostered an incoming glut of fissile material that required secure transportation and adequate storage capacity. Yet, during Nunn-Lugar 's first couple of years, MC&A received comparatively little attention.

Concern grew within the United States and in western Europe through the early 1990s that nuclear material such as plutonium and HEU from dismantled nuclear weapons or nuclear research sites in the FSU might find their way into the hands of terrorists or rogue states. Widespread reports of nuclear smuggling in 1993-94 further compounded the growing perception of the threat of nuclear leakage to the United States. Together, these questioned whether the United States was doing enough in the area of MPC&A.

Thus, political pressure mounted in Washington for the administration to place more emphasis on fissile MPC&A Accordingly, formal U.S.- Russian government-to-government cooperation on such issues began with a modest initial $10 million Nunn-Lugar outlay in 1993. Yet, despite the earmarking of funds in successive pieces of legislation, progress on MC&A efforts such as the fissile material storage facility appeared hopelessly slow, and the obligation rate of funds fared about as it did in other DOD programs. At the same time, an influential study issued by the National Academy of Sciences called the problem "urgent" and referred to it as a "clear and present danger" to U.S. national security. By January 1994 President Clinton and Russian President Boris Yeltsin had declared the risk of nuclear theft a high priority, and high-level attention was sustained in successive presidential summits and meetings of the Gore-Chernomyrdin Commission (GCC) through at least 1996.

Efforts designed to combat nuclear leakage thus represented a natural evolution of programmatic needs that focused sequential attention on destruction and dismantlement increasingly toward chain of custody issues. Under the leadership of Secretary O'Leary, Deputy Secretary of Energy Charles Curtis, and NN-40 Director Luongo, DOE advocacy of a variety of policies and the tactics to implement them differed increasingly from the State Department-led Safeguards, Transparency, and Irreversibility (STI) talks headed by Ambassador Goodby, as well as other related efforts. While State retained the diplomatic lead on government-to-government MPC&A discussions with Russia, the "irritated relationship" between DOD, State, and Minatom revolved around the perceived zero-sum underpinning of this formal arms-control approach and allowed only minor progress in this area. Thus, even as Ambassador Goodby was tasked by the vice president's office with conducting these high-level negotiations (and received support from OSTP and some State and NSC officials), senior officials in DOE and DOD, together with key NSC staff, increasingly made the case that formalized, high-level involvement in MPC&A issues would actually serve to stymie U.S.-Russian progress in this area.

According to Luongo, by 1994 this acted as a catalyst for a "frustrated" DOE to begin in earnest a parallel laboratory-to-laboratory MPC&A approach that circumvented national-level negotiations. While DOE faced an "uphill battle" in gaining support for such bottom-up MPC&A cooperation against a resistant DOD, State, and some NSC staff, it found allies in OSTP, elsewhere at NSC, among some members of Congress, some Russian nuclear installations, the U.S. Embassy in Moscow, and, after some time, the Office of the Vice President. While the NSC did not provide effective leadership on this matter, State was uncomfortable with the more informal nature of this approach and DOD refused to provide any financial support for the enterprise for several months. Indeed, only after the General Accounting Office (GAO) and PCAST issued in late 1994 and early 1995 studies that were critical of the U.S. response to the fissile material security situation in the NIS did DOD officials agree to provide DOE the $15 million it sought for laboratory-to-laboratory MPC&A cooperation in 1995.

In the context of a shrinking budget approximately $19 billion in 1993 and $16 billion in 1997 DOE planned to request at least $800 million in support of its planning for nuclear material security in the NIS between FY 1996-2002. It estimated the initial cost of establishing an effective Russian MPC&A program at more than $1 billion. While DOD worked with Russia's MOD and remained committed to facilitating warhead security, DOE committed to a "Partnership for Nuclear Material Security" with Minatom and the Russian civilian regulatory agency, Gosatomnadzor (GAN).

While the government-level dialogue on cooperation "quite understandably, reposes with the Department of State, " the genesis and development of the laboratory-to-laboratory MPC&A program was a useful institutional innovation. While largely using its own political and financial resources to create cooperative efforts between U.S. national laboratories and Russian institutes, by the end of 1996 DOE directly helped to raise the fissile material security standards of forty-four Russian nuclear facilities (approximately 80 percent of the total) where weapons-usable materials are stored. This was in part due to looser, more results-oriented contracting regulations that expedited the flow of funds well beyond what standard DOD contracting procedures would allow. Other programs developed by DOE, such as the Industrial Partnership Program (IPP, later renamed the Initiatives for Proliferation

Prevention) drew a consistent base of support from members of Congress such as Senator Pete Domenici (R-NM), even as others on Capitol Hill and within DOE headquarters attacked the program each year after its 1993 creation (see chapters 4 and 6). As a result of DOE 's efforts and DOD's preference to relinquish control over many of these programs, all programmatic and funding authority for the MPC&A and IPP programs were formally transferred to DOE as that agency became their executive agent in FY 1996. DOD formally retained jurisdiction over, but continued to work closely with, DOE on warhead security, fissile material storage, and many other issues of common concern.

Department of State

In contrast to DOE, State played a strong leadership role in CTR from the program's inception. State Department involvement in Nunn-Lugar started at the highest level, with Secretary Baker (and, to a lesser extent, Deputy Secretary of State Lawrence Eagleburger) taking a personal interest in U.S. policy toward a rapidly disintegrating Soviet Union. Other prominent officials such as Ambassadors Robert Gallucci and Richard Armitage were also deeply involved in early post-Soviet US-NIS relations, the former with the science centers and the latter with humanitarian assistance efforts. Although Secretary of State Warren Christopher played a lesser direct role in U.S. regional policy toward the NIS, he did speak widely on the subject, testified repeatedly before Congress, and was involved in the annual CTR certification process. More importantly, Strobe Talbott, as deputy secretary of state and before that as ambassador-at-large for the NIS, devoted considerable time and attention to regional policy setting. While waning considerably after 1994, sustained, high-level advocacy characterized State's involvement in Nunn-Lugar during both the late-Bush and early-Clinton administrations.

The initial authorizing legislation made DOD the program's executive agent on the premise that this was a time-sensitive defense initiative that should not be delayed by the usual, lengthy arms-control process. Senator Nunn argued that Congress "did not lightly put in there that we expected DOD to be the executive agent. That was done with some forethought." Noting that State deals in "diplomatic areas" rather than in technical implementation, he explained his committee's intent as follows:

And what I fear is we have a State Department in control over here. The Soviets feel like they've got to have their State Department people in control, and before long you have been negotiating for nine and ten years on every arms control agreement when we really do have some urgency here. Arms control negotiations have their place, but I don't view this as a negotiation that should be lasting for a matter of years.

As a diplomatic rather than technical implementing agency, State clearly did not have the financial or logistical resources of DOD. Indeed, given Nunn's remarks, it seems clear that Congress did not intend for State to play such a large role. But while DOD was indeed the executive agent by law, in practice State took the diplomatic lead early on.

Nevertheless, the departments apparently thought along similar programmatic lines during much of CTR's first two years. In accordance with the Bush-Gorbachev reciprocal unilateral arms control initiatives of September 1991, and in conjunction with START I requirements, early Nunn-Lugar efforts emphasized measures to ensure the safety and security of nuclear weapons primarily during their transportation and storage. According to Bush administration Deputy Assistant Secretary of State Robert Walpole, senior State Department negotiators pressed for early agreements in the following five areas: (1) storage containers for fissile material and storage facilities; (2) armored blankets and safe and secure railcars for transport; (3) nuclear weapon accident response equipment; (4) the development of a system of fissile material control; and (5) the ultimate disposition of HEU and plutonium from dismantled weapons.

While the latter two items were of a decidedly long-term nature and did not find an immediate and widespread advocacy, the first three found support among both DOD and NIS officials. Both State and Defense, along with their Russian counterparts, quickly agreed that the most urgent problems they faced were the transportation and storage of the still-dispersed Soviet tactical nuclear arsenal, due mainly to the vulnerability of such assets in terms of control, loss, or unauthorized seizure. But by late 1993, and especially by late 1994,

DOD and State policy preferences began to diverge considerably. Clinton's DOD team strongly emphasized NIS denuclearization even as they pursued broad-based demilitarization objectives. At the same time, State Department officials attempted to retain their active leadership role in CTR, including a greater emphasis on nonproliferation tasks. This led to "big fights" with DOD officials over money, even as State-DOE tactical disagreements on policy implementation flared up. Increasingly, senior DOD and DOE officials disputed State's efforts to formalize the laboratory-to-laboratory process, fearing that too much high-level involvement (often derided as the "old think" of arms-control-business-as-usual) would ultimately jeopardize rapid progress in Russian MPC&A improvements.

Program Management and Coordination

There appear to have been few real internal conflicts at Foggy Bottom. The Bureau of Political/Military Affairs (PM) and, more specifically, its Office of Strategic Policy and Negotiations, took the lead in programmatic and budgetary coordination at the working level. PM's role was to coordinate the State Department's involvement, as well as support U.S. ambassadors in the NIS and administer the science centers in Kiev and Moscow.

State had both direct and indirect influence on the Nunn-Lugar program. Regarding the former, not only was State directly represented in the Nunn-Lugar IWG, but in FY 1996 it also became responsible for the programmatic and financial aspects of the science centers in Kiev and Moscow. Moreover, it helped establish or modernize NIS export control policies, it maintained the formal diplomatic lead on government-to-government MPC&A cooperation, and it provided modest financial support for DOD's program of military and defense contacts in FY 1993-94. Thus, other offices within PM, such as those dealing with export controls and the science centers, played specialized supporting roles.

The more informal mechanisms for State's involvement in Nunn-Lugar stemmed mainly from the department's organizational relationship with the ambassadors who led the SSD negotiations from January 1992 through March 1994. Even though their status as special representative and chief negotiator of the president for NIS denuclearization gave them a direct link to the White House, they were physically located in the State Department. Their location, along with requirements to report directly to State leadership Major General Burns reported to the Under Secretary and Ambassador Goodby to the Assistant Secretary for Political/Military Affairs facilitated a clear State lead on Nunn-Lugar negotiations, a structure made possible in part by general resistance to CTR efforts by the Cheney Pentagon.

The structure and composition of the SSD delegations was reminiscent of prior arms control negotiating teams. General Burns, SSD ambassador under Bush, had two deputies: Assistant to the Secretary of Defense for Atomic Energy John Barker and Energy Department Assistant Secretary James Turner. The three principal agencies involved in the SSD efforts were thus represented from the very beginning of the program. State acquired the nominal lead on diplomatic efforts, and Defense and Energy were included because they each had institutional equities to preserve or particular roles in policy setting or implementation. When Ambassador Goodby replaced Burns, he also had two deputies: DOE's Turner and Assistant to the Secretary of Defense for Atomic Energy John Birely, who was later replaced by Gloria Duffy as special coordinator on the policy side of OSD.

In addition to the SSD ambassadors, the Office of the Special Coordinator for Assistance to the NIS (S/NIS/C) was also physically located within State, although Ambassador Richard Morningstar had direct reporting responsibilities to the national security advisor. Headed briefly by Ambassador Talbott before he became deputy secretary of state, this office was also formerly headed by Ambassador Thomas Simons, who resigned in April 1995. The special coordinator was charged with overseeing all economic aid and bilateral assistance programs relating to the former Soviet region and often acted as the administration's principal advocate of NIS programs to Congress.

The NIS special coordinator was chartered under the Freedom Support Act of 1992, in part to help resolve policy and program disputes among the various U.S. government agencies involved in NIS assistance programs. The initial May 1993 charter was signed by Anthony Lake, President Clinton's national security advisor, and required that the coordinator report to both the ambassador-at-large at State and a special assistant to the president. But, as a February 1995 GAO report observed, the coordinator had "limited or no authority to direct activities of the Cooperative Threat Reduction program, " and, thus, had "no way of ensuring that all programs for the FSU complement one another." The special coordinator was not a regular member of the NSC-led Nunn-Lugar working group and, at least until April 1995, had no formal authority over CTR funds. GAO concluded in its report that the coordinator came to the interagency bargaining table from a "weak position, " finding his authority "frequently challenged" by intra-executive bickering, congressional earmarking, and negotiations undertaken by the Gore-Chernomyrdin Commission.

In March 1995 testimony before the House Committee on International Relations, Ambassador Simons responded to Congressman Lee Hamilton's (D-IN) concern that "the reason the program has been so slow to get along has been because the coordinator has not had sufficient authority, " by noting that "there will always be disagreements among agencies when implementing a program of this size, importance, and complexity." Nevertheless, President Clinton signed an expanded charter for the new special coordinator, Richard Morningstar, in April 1995. This newly expanded charter mandated that the coordinator preside over the allocation of U.S. assistance resources and direct and coordinate the interagency process on the development, funding, and implementation of all U.S. Government bilateral assistance and trade and investment programs related to the NIS.

Thus, the coordinator gained at least formal coordination responsibility for the nineteen U.S. government agencies involved in NIS assistance efforts, including DOD.

According to officials within S/NIS/C, this expansion of the special coordinator's charter gave Morningstar jurisdiction over the Nunn-Lugar program. White House attempts to expand his portfolio and provide him with a higher profile indicated at least a tacit admission that interagency coordination was not as smooth as it could have been or, in the face of increasing criticism of NIS aid by the Republican majority on Capitol Hill, needed to be. At the time, others voiced the opinion that this merely made Morningstar a "paper tiger, " that the NSC still had the coordination lead while DOD maintained primary day-to-day program implementation responsibility.

GAO reexamined the coordinator's role later in 1995 and found that in most areas executive-level participants agreed that policy coordination had improved, that his role had been strengthened, and that the working relationship between the various implementing agencies had improved. Nevertheless, the lines of authority between the special coordinator and the NSC working group on Nunn-Lugar remained unclear: Coordination between S/NIS/C and the NSC appeared to rely more on personal relations than formally designated authority. Indeed, Morningstar and Coit Blacker, the NSC Senior Director for Russia, Ukraine, and Eurasia, each held the title of special advisor and were of comparable rank. Nevertheless, since most of the policy (as opposed to implementation) issues already had been settled, the coordinator's new-found authority over CTR would never be adequately tested.

Summary

DOD planners consistently ranked strategic denuclearization of the non-Russian former Soviet republics as priority number one, followed by assistance to Russia to help it comply with the terms of START I. Defense conversion and demilitarization were also strong DOD priorities. The Energy Department and OSTP, with considerable State Department and NSC support, gradually pressed for increased attention to MPC&A and nonproliferation objectives over defense conversion, demilitarization, and even dismantlement efforts. While they shared DOD's core objectives, they did not rank them as highly. But in light of DOD's considerable financial endowment and implementation responsibility, it quickly became the central actor and preserved many of its CTR policy priorities.

In 1994-95, senior DOD officials brokered the fragmentation (or balkanization) of the Nunn-Lugar program into its constituent elements. This was considered by some policy officials, especially NSC Director Gottemoeller and SSD Ambassador Goodby, to be a mistake since it invited uncoordinated policy efforts that risked reducing overall U.S. leverage or otherwise diminishing Nunn-Lugar 's influence. While their charge proved accurate from a coordination standpoint, from a resource perspective their fears proved to be largely misplaced. By fostering State and DOE institutional lobbies, program management and focus improved and ultimately facilitated a larger net Nunn-Lugar program budget than would otherwise have been possible in the face of severe congressional criticism. Indeed, the essential tension between programmatic fragmentation and policy centralization yielded two primary results: (1) a substantial increase in resource allocation for threat-reduction efforts spread among DOD, State, and Energy beginning in 1996 and (2) a notable lack of policy coordination among these central actors, who pursued them with considerable autonomy.

7

Strategic and Political Implications of Missiles

Introduction

The future has potential implications of such magnitude and variety that it is virtually impossible to assess them accurately. Indeed, missiles are weapons of such incredible speed and destructiveness that their influence extends throughout the whole empire of military affairs and beyond to the international politics of peace and war. Never before have events moved so rapidly ahead in the development of engines of destruction that mankind has been compelled to doubt its ability to halt the rush toward world-wide annihilation. Yet, thus for the implications of missiles have been largely left for the military expert and the politician to weigh. It is now time for men of different perspective to study them, for survival depends upon bringing as wide a spectrum of ideas as possible to bear on the problems which confront us. Only in this way can the nation react intelligently and shape its course to meet the coming challenge.

The missile age is now well toward the end of its second decade if we date its birth at Germany's success with V-2 rockets in World War II. But, for the average American, it only came alive late in 1957 with the dramatic announcement that the Soviet Union had placed a satellite in orbit around the earth. Initial reaction to this feat was one of incredulity plus a strange mixture compounded of both concern and unconcern. To the American people, it was almost unbelievable that Communist Russia could have bested the United States in a technological contest, for Americans had long taken their own invincibility for granted in matters requiring high scientific and industrial skills. To a nation accustomed to technological leadership, and equally accustomed to measuring strength by this standard, the tangible and shocking evidence of Soviet superiority created grave concern for the national security. This was all to the good. Fortunately, the Soviet Union had demonstrated its strength in such a way that a resolve to compete rather than a sense of resignation took hold of the American people. If there was anything they could do to provide money, brains, or resources, national leadership needed only to ask for it.

Curiously, however, the reaction among many leaders who might have been expected to know better was one of seeming unconcern. The Soviet satellite was shrugged off as "a silly bauble" a "neat scientific trick." This disturbing attitude was undoubtedly based in part on a lack of appreciation of the true significance of the Soviet accomplishment. The satellite itself did not appear to have any immediate or vital military significance, but the engine which boosted such a weight of metal beyond the earth's atmosphere showed that the Russians were well advanced in rocketry. By the time this fact had been spelled out, and the passage of time had permitted the domestic political repercussions of this Soviet victory to die a natural death, there were few men left in Washington who did not fully understand that the United States was in a race with the Soviet Union which could easily end in destruction for the loser. As a result, an interest in missiles has finally been kindled which is likely to produce substantial accomplishments. If our efforts are to be channeled in the right direction, however, there must be a clear understanding of just what the influence of missiles on strategy may be.

The succession of satellite launchings from both sides of the iron curtain and the test-firing of prototype, long-range ballistic missiles by the United States have focused attention on the more spectacular and newsworthy members of the missile family. But no appraisal can be complete or realistic without considering the full gamut of instruments which are coming into existence. The giant of the family is, of course, the intercontinental ballistic missile, the ICBM, with a range of more than 5,000 miles. Scarcely less important for the United States, with its numerous overseas bases, is the intermediate-range ballistic missile, or IRBM, which can reach targets 1,500 miles away. These two may be classified as the strategic weapons of future military arsenals, comparable to strategic bombers of the present, for they are offensive weapons pure and simple, capable only of mass nuclear destruction. They are too expensive to waste on conventional warheads and too inaccurate for precision firing at pinpoint targets. It is their strategic implications and the knowledge that at the moment there is no defense against them that have given them such a prominent place in the public eye.

The pygmies among the missile family are, however, of equal importance. They are much more numerous; and many are in actual operational existence. Among these are types which have capacities for defense and limited war that are already making themselves felt. Far from being restricted to strategic operations, they have implications for tactical warfare on the ground, in the air, and at sea which give them weight out of all proportion to the attention being given to them. Moreover, it is in this category that defense against the long-range missile may be found. If it were not for the short-range missile, we could expect to live out the balance of our existence under a reign of terror, awaiting whatever madman pushed the button to exterminate us all. It is the possibility of developing an effective anti-missile missile that lends hope for the future of the human race.

Perhaps the most difficult task in assessing the implications of missiles arises from the rapidity with which new weapons are being developed, and from the uncertainty over how the balance between offense and defense will shift in the years ahead. The chronology of development is so unpredictable that virtually any statement made or any strategic planning done today may have to be reversed tomorrow, and the conflict between current, mid-range, and long-distance planning is keeping strategists and policy-makers in a constant turmoil of indecision. Already, some of the weapons on which millions of dollars and months of time have been expended have been rendered obsolete by the astonishingly quick development of some other weapons. The problem of selecting which types of weapons to advance and which to curtail becomes increasingly difficult as costs mount and as the consequences of erroneous decisions become crucial.

This question of allocating research funds or freezing production on one kind of weapon as opposed to another has, of course, existed for many years, but never has it been so acute as now. It is tempting to assume, for example, that the missile will entirely replace the manned bomber, or that any future war must be terminated in a matter of hours, so that there is no need to build elaborate defenses or to stockpile human and material resources as a strategic reserve, or even to create forces for limited war. But none of these things is entirely or inevitably the case. There is no question that missiles are raising some problems which differ radically from those we have known before. Yet, in some ways the introduction of missiles is merely intensifying and making more acute the problems we already face from the existence of nuclear bombs and long-range bombers in the hands of the enemy. Some of the old problems, in other words, remain with us, but they are of quite a different order of magnitude. The problem of base mobility, for instance, becomes increasingly important in the missile age. The roles and missions of the services become subjects of long dispute, requiring closer and more constant examination than ever before; and questions of personnel policy and of tactical organization are also requiring rapid readjustment.

It is the sharply increased velocity of the missile over the airplane which, more than any other single characteristic, is intensifying old problems and causing military experts to recast some of their most modern concepts of military policy and strategy. Prior to World War II, for example, the limited range of weapons, combined with the relatively long time span necessary for deploying overwhelming force, permitted nations considerable latitude in the composition and readiness of their military forces. The United States particularly, isolated by its unique geographic position, was always able to get by with the shadow of a professional army until after the outbreak of war. It could even afford to ship a considerable proportion of its scarce military equipment to Great Britain in 1940 and 1941, denuding its own meager military establishment in the process, without incurring undue risk to its security. In the decade following World War II, however, the development of massively destructive atomic and thermonuclear bombs, coupled with the intercontinental delivery capacity of piloted aircraft, so compressed the time scale of warfare that old concepts of readiness became outmoded. As a result, the traditional American military policy of depending on a small professional force, supplemented by large numbers of half-trained reserves and hordes of unskilled citizen-soldiers, became anachronistic, and a new concept of military strength measured in terms of forces in being came into existence for the first time.

The advent of missiles has greatly intensified the compression of the time scale for modern warfare which the intercontinental bomber brought into being such a short time ago. Both the United States, protected by oceans, and the U.S.S.R., protected by vast land masses, have had to adjust radically to the new situation. Missiles, as they come into the final state of their ultimate potential, will increase the whole tempo of war. Not only will surprise attack be made easier and defense more difficult, but the old soldier will have to be replaced by the electronics expert. When missiles can approach their target at 10,000 miles an hour, a problem which seemed difficult enough at the time the long-range bomber came into existence will wrench the defense out of human hands onto the electronic maneuver board.

If there is to be any defense against such weapons, it will have to be entrusted to machines, for the reaction time will be beyond the capacity of the human brain. From time of launching to the instant of impact a matter of moments at best the missile will have to be identified as real or decoy; its trajectory will have to be determined, and its path to the target tracked constantly. And all this will have to be done instantaneously if the opposing missile is to intercept it in flight. For the ICBM, perhaps thirty minutes will be available. For the IRBM from perhaps five to ten minutes will be available. But on battlefields of the future, where short range missiles will be employed, bare seconds will be allotted for all these calculations. To say that the tempo of war will be speeded up is to say that our whole thinking with respect to readiness and reaction time must be drastically altered.

It is apparent, therefore, that national security in the future will rest not only upon forces in being but on types of men and equipment that can react in terms of fractions of time as compared to previous requirements. No matter how efficient missile detections systems become, this reaction time will be measured in minutes or even seconds, and this applies to the defense of our production facilities and population centers, our fixed bases and the retaliatory forces operating from them, and the forces we have deployed on foreign battlefields as well.

The Strategic Air Command, at present our principal force for total nuclear war, for example, adopted an alert system some time ago requiring one-third of its total bomber complement, combat loaded, to be airborne within fifteen minutes, twenty-four hours a day and seven days a week. This may have been ample for the age of manned aircraft, but, with existing detection systems, it will not be good enough to counter the ICBM. Moreover, it is not sufficiently fast to counter the IRBM launched from enemy submarines with any detection system that may be devised, because the total flight time of such missiles will be less than ten minutes. If the assumption is that the United States will not initiate hostilities but only counter assault by others, it must be prepared to react instantaneously. Its forces will have to be capable of delivering decisive retaliatory blows against any potential enemy in the face of surprise attacks so swift that manned bombers will have no chance to become airborne before they are destroyed. To insure this capability, and thus deter attack in the first instance, those of our retaliatory forces whose positions can be predetermined by the enemy will have to be able to react in five minutes or less, even if instantaneous detection systems can be developed. Such fast reaction time is not feasible with current piloted aircraft operating from huge immobile air bases, nor will it be possible with the liquid fueled and highly complex long-range missiles that are scheduled to become operational in the near future.

The solution to this difficult problem of readiness and reaction time, therefore, does not lie solely in superior detection systems, although these will be essential, nor does it rest with manned aircraft, which cannot respond fast enough to clear their stationary bases and escape destruction. The problem is a twofold one requiring, first, the development of short-range, solid-propellent antimissile missiles capable of instantaneous launching to strike at the weapons which can destroy our civilization, and, second, the creation of mobile launching sites from which our own long-range missiles can deliver counterattacks upon the enemy. In other words, the solution lies in a combination of the defense and offense, for only so can we insure our own survival and security, and compel the enemy to be cautious about risking his own.

The influence of missiles on the offensive-defensive balance is particularly difficult to assess. Whatever is said this year with assurance about the balance will undoubtedly be invalidated next year, perhaps to be restored to credibility a few years hence. The pendulum between the offense and defense in military operations is, of course, forever in motion, but the swift development of missiles has speeded up its movement to the point where it is almost impossible for strategic planning to keep pace.

Only a few years ago, for example, with the advent of intercontinental bombers carrying nuclear payloads, the offensive swiftly outpaced the defensive because interceptor planes could not be certain to prevent a considerable fraction of an attacking force from penetrating to its targets, and a small force was enough to do untold damage. This prompted the strategy of massive retaliation as the only means of countering an enemy attack. Today, however, the balance is being dramatically altered by reason of the fact that an imposing array of short-range missiles, which seem capable of dealing effectively with the manned bomber, has been introduced into the forces of the principal antagonists. Yet, despite the fact that SAC bombers will find it increasingly difficult to penetrate to the vitals of the Soviet Union, and despite the fact that our own defenses are better able to turn back an enemy attack, the strategy of instant retaliation by long-range bombers continues to maintain its grim stranglehold on our strategic thinking.

When ICBM's and IRBM's become operational in sufficient quantity to be of military significance, the offensive may be restored to the place it held a few years ago, and a retaliatory doctrine, based this time upon missiles, may once again attain a certain validity. But that time has not yet arrived. In spite of the success recently experienced with prototypes, and despite the accelerated programs in evidence on both sides of the iron curtain, it will be several years before ICBM's and IRBM's will restore the offensive to its former high position. The significant fact today is that short range missiles are shifting the balance in favor of the defensive, and yet billions of dollars are still being poured into a manned bomber force which is becoming increasingly outmoded. Moreover, there is every evidence that the U.S.S.R is going to beat us to the punch in the field of ICBM's and will have these weapons operational before we do. Unfortunately, efforts to begin work on an effective anti-missile missile program were long stultified in the United States by an attitude which either failed to see the imperatives of the future or dismissed the idea as not worth trying. This now makes it doubly important that highest priority should not be going to SAC but into the missile and anti-missile programs which are desperately needed.

While strategy must be based on the performance of weapons currently in existence, research and development policies have to be based on the types of performance needed in the future. It may well have been a failure to understand this distinction which caused the United States to lag in its weapons planning. Today total war would be a contest between the short-range missile and the long range bomber, with the missile tipping the balance rather firmly in the direction of the defensive. This is primarily due to the fact that missiles can strike unerringly at manned bombers despite the latter's supersonic speed and relatively high ceiling. Piloted aircraft simply cannot compete with the missile either in rate of acceleration or in ability to escape from the atmosphere. Air-to-air missiles are of much greater value to the interceptor than they are to the bomber and, when combined with surface-to-air missiles, they add real substance to air defense. They are already capable of exacting a heavy toll on attacking aircraft and can be expected to curtail strategic bombing still further as future refinements make them more accurate and reliable. In other words, the manned bomber is already facing obsolescence, not from the ICBM's of the future, but from the short-range missiles of today.

It may well be at this point that the United States made its error. It was apparently assumed by those in supreme authority in the Department of Defense that the piloted bomber would not become obsolete until the long-range missile became available, whereas the fact is quite the contrary. It is the short-range missile which is making the bomber obsolete, and it is this in turn which has made necessary the development of intercontinental missiles as replacements for the bomber. The U.S.S.R. apparently understood this chronology better than did the United States, and it began its long-range missile program early, knowing the defensive missile would soon outstrip the offensive bomber.

The influence of the short-range missile on the offensive defensive balance also has serious implications for the ground forces, where extremely effective anti-tank missiles are drastically reducing the effectiveness of armor, the soldier's principal offensive arm. Indeed, there is little question that the introduction of a variety of tactical surface-to-surface missiles as replacements for conventional artillery is going to bring about radical changes in the whole tactical organization of army units. The high accuracy of the command guidance system, which is making possible direct hits in four out of five cases with surface-to-surface missiles of 55-mile range, has already made the cost of tank destruction relatively low. It is an open question whether this development outweighs the added firepower and versatility of other missiles employed by ground forces in offensive roles, but certainly the United States and its NATO allies are counting on nuclear fire- power and short-range missiles to stop the onslaught of Soviet armies, if they should attempt a headlong rush across Western Europe.

The Soviet Union has, however, apparently developed a midrange ballistic missile with a capacity of perhaps 450 to 700 miles, and with this deployed in numbers in support of its advancing troops, the defense may well find itself overwhelmed. The United States Army has no missile overseas that can compete with this in range or in offensive capacity, nor has it any anti-missile that can defend against it. The anti-missile program has been long delayed through decisions that have kept it short of funds, and the Army's mid-range missile program was cut short, at the end of 1956, by a Defense Department directive that limited the range of its tactical surface-to-surface missiles to 200 miles. These decisions were presumably prompted by a strategic concept which discounted the availability or offensive capabilities of the Soviet mid-range missile, or which preferred to look farther forward to a time when our own IRBM's could be counted on to outweigh the effectiveness of the mid-range weapon.

However the balance swings in ground operations or in limited wars in the next few years with the development of tactical missiles, there is no doubt that the time will come when intercontinental and intermediate-range ballistic missiles will give the offense a clear edge over the defense in total war. What now seems to be a trend in the direction of the defensive, because manned bombers cannot compete against defensive missiles, will surely be reversed when long-range missiles can be launched to strike at the vitals of the belligerents' homelands. There are those who contend that when this point is reached, when the marriage of long-range missiles with thermonuclear warheads is consummated, the pendulum will have swung full-scale to the offensive, there to remain. This is the basis for the continuing strength of massive retaliation as the basic strategic concept of the United States an assumption which may, however, well lead us into a dangerous rigidity of thought, for, potent though the long-range missile with thermonuclear warhead may be, it is still not the ultimate weapon.

Even before the ICBM and IRBM have become operational, work has been started on defenses against them, and chinks have begun to appear in their supposedly impenetrable armor. There are experts in this field who are convinced that anti-missile missiles are feasible and who confidently predict their swift development and deployment. Their confidence is based upon the fact that the ballistic missile must follow a predictable trajectory and that, once the path is charted electronically, the anti-missile missile can be set to intercept it. ICBM's will, of course, be less of a problem than the IRBM. They will be in flight longer; and their trajectories between Soviet Bloc areas and the United States are extremely inflexible. They will approach their targets from a nearly vertical angle at a predictable speed, and the defense system can be preset. In fact, there is reason to believe that missile defense against Soviet ICBM's can be provided by the early 1960's if development of the Army's Nike-Zeus is given high priority and pushed with sufficient funds and sufficient vigor.

Defense against the IRBM is, however, a different matter. The IRBM will be in the air for a much shorter period, perhaps five as opposed to thirty minutes, and its trajectory will be more variable, thus complicating the problem of defense. A defense sufficiently deep to prevent penetration of the missile launcher to firing range is, of course, the simplest solution to the problem, and this the United States fortunately can provide for itself upon the seas or across the polar cap. Western Europe and other areas contiguous to the Communist orbit cannot. Time and space factors make it imperative for them to seek their defense solution in the development of anti-missile missiles of such instantaneous response that they can strike their targets while in flight a problem which, though not hopeless, will certainly remain unsolved until long after IRBM's have been deployed against such areas in quantity.

Probably the most significant point about this ever-changing contest between the offense and the defense, and one that is frequently overlooked in our emphasis on the strategy of retaliation, is that only in a clear-cut victory for the defensive can the world be sure of survival. Long-range ballistic missiles are weapons of such speed and power that both sides will be utterly destroyed in a total war unless air defense through anti-missile missiles can be made reasonably effective. Without air defense, the best that the doctrine of retaliation can do is to improve the chance of deterring attack by promising the destruction of both belligerents. It is always possible, however, that such a war might be precipitated by accident or by design based on miscalculation. The ballistic missile, once launched, cannot be recalled. In such an event, the side that is able to come closest to a balance of offensive power and defensive strength will stand the best chance of survival. It is defensive capability, as much as retaliatory capacity, that holds the key to national security, and this must be kept in mind in the development of future weapons and strategies.

True as it is that strategic thinkers must constantly keep an eye cocked to the weapons of the distant future, concrete strategic planning cannot, nevertheless, be based upon such remote factors. Today's problem, therefore, is to take account of the increasing obsolescence of the manned bomber, the expectation that long range missiles will soon become available in quantity, and the fact that there will be no effective defense against them in the immediate future. It is these three factors that will give the doctrine of retaliation renewed validity in the next few years and that make it so essential that we look for new means to keep our offensive capacities high. As long as we lack defenses against long-range missiles, large fixed bases for our retaliatory weapons will remain subject to swift destruction, and our attention must be turned to the development of mobile bases for our own long-range offensive instruments of retaliation. It is this necessity for base mobility that has created the current debate about sea-based versus land-based missiles a debate which has so far been conducted mainly by naval spokesmen on the one hand and by those of the Air Force on the other.

The Army, with a role which is primarily tactical in nature, knows full well the necessity of mobility for its short-range missiles, and its efforts have been unceasingly in this direction. It learned this lesson during World War II, when Germany's fixed, concrete missile bases were easily destroyed by conventional bombing while not a single one of her mobile missile units was ever detected until Allied ground forces overran them toward the end of the war. The Army knows also that the larger the missile the more difficult it is to move it readily. The huge ICBM's, which must be able to respond instantly and accurately to a Soviet attack, are virtually impossible to divorce from fixed bases, while the Army's missiles are short- or mid-range weapons employed for tactical purposes, and as such do not come into direct competition with the strategic missiles of the Air Force and Navy. It is over the strategic missiles that the current dispute is raging.

The Navy's case for base mobility is founded primarily upon the advantages which can be derived from a marriage of the IRBM as a missile and the sea as the medium from which to launch it. It is within the realm of possibility that the IRBM launched from ships at sea can achieve the same target coverage as the land-based ICBM. Two geographical factors make this a matter of great potential significance. In the first place, missiles with a range of 1,700 miles could, if launched from the proper points at sea, reach any land target on earth. Moreover, because more than half the world's population, plus the industrial complex which supports it, is located within fifty miles of the sea, shorter-range weapons based on ships far at sea could achieve sufficient destruction to cripple effectively any nation which they struck. The weapon on which the Navy is basing its present claims is the Polaris missile, an intermediate-range weapon which has been brought to the brink of operational readiness far in advance of schedule. Initial estimates placed it well behind similar land-based missiles in regard to production date, but it has already reached a stage of development which entitles it to consideration in current strategic plans.

Among the advantages which are claimed for sea-based missiles by spokesmen of the Navy are these: greater accuracy than the ICBM and elimination of the problems involved in maintaining bases for IRBM 's on foreign territory. The latter is a political problem which will be discussed later, but the claim of greater accuracy is essentially military in its implications. It is based upon the proposition that there is an inverse relationship between range and precision. Land- and sea-based IRBM 's are supposed to have equivalent characteristics of accuracy, but the sea-based Polaris with shorter range should be more of a precision instrument than the intercontinental ballistic missile launched all the way from the United States. Yet, it is not clear whether this means that the IRBM can be counted on to take out the enemy's fixed bases, while the ICBM cannot. If this is not the case, it will have little relevancy unless indiscriminate area bombing is to be ruled out in strategic warfare of the future. The use of IRBM 's, whether launched from land or sea, is most unlikely in tactical operations or limited conflicts. They are strategic weapons pure and simple.

The principal advantage of sea-based missile systems, however, rests upon their mobility. From this mobility they derive their relative invulnerability and the assurance that they will survive to launch a retaliatory attack. Being mobile, they can be assured of immunity from attack by enemy ballistic missiles, for the latter must be aimed in advance of firing, and thus cannot be employed successfully against moving targets. Moreover, the nuclear-powered submarines which will launch Polaris missiles, being hidden in the vastness of the area beneath the surface of the seas, beyond the reach of either visual or radar reconnaissance, will be most difficult to locate and identify. There is little question, then, that the IRBM, when harnessed to the submarine, will add tremendously to the invulnerability of our retaliatory forces by the ease with which it can escape detection and destruction an ability which is not shared by the easily identifiable, fixed launching site of the ICBM.

This is a sufficient reason in itself to incorporate Polaris-type submarines into our retaliatory system. But Navy spokesmen like to carry their case one step further by suggesting that these submarines will act as lures for drawing much of the enemy's attack away from our homeland and toward our retaliatory forces at sea. Every missile and every force expended against Polaris launchers, so goes the argument, is a missile or force subtracted from the potential suffering of our people. Land-based missiles within our borders will, on the contrary, only attract greater destructiveness to us. This is an appealing argument, but it is a little like trying to have the best of both worlds. If mobile undersea forces are as difficult to locate and as invulnerable to enemy ballistic missiles as they appear to be, it is unlikely that the Russians will waste their initial strikes upon them. Surface forces carrying ballistic missiles, being easier to locate, may have this magnetic attraction, but it is unlikely that the enemy will use fixed-trajectory missiles against them either.

The introduction of the Polaris-type submarine into the naval arsenals of the world creates a grave problem of assessment for our own national security. The Soviet Union apparently recognizes the relative invulnerability and the striking power of this weapons system, and is pushing ahead with its development with all possible speed. Already, it has hundreds of conventional submarines, and, when it adds a few missile-firing, nuclear-powered submersibles to its underseas fleet, the Soviet Union will have presented the U.S. with an extraordinarily difficult problem of defense. The conventional submarine is basically a tactical weapon: only the missile submarine has capacities for strategic offense. But the sheer number of underseas boats which the Soviets can keep at sea will give the United States Navy a prodigious task of location, identification, and attack upon them. Since the United States does have a far more powerful defensive surface force available to locate and attack underseas craft than do the Russians, the scales of strategic warfare may be tipped in our favor. But this will not be accomplished without a major effort, and certainly the development of higher quality and quantity in the anti-submarine forces must be given a high priority in our military scheme of things.

Defense against the nuclear-powered submarine, difficult as it is, is probably easier than attempting to intercept its missiles in flight. One of the advantages of the sea-based missile, in comparison with the land-based ICBM, is its shorter flight time, which gives it greater surprise value and complicates the problem of the defender. Moreover, sea-based missiles launched from a great variety of angles, add to the defender's problem of detection and compel him to make an increased defense effort and a greater expenditure of funds and resources.

The question of cost is important, but whether the ICBM or the Polaris system has the advantage on this point is almost impossible to determine. With an equally powerful warhead but a smaller thrust mechanism, Polaris undercuts the ICBM in cost of manufacture, but when the missile ship is added to the expense of the sea-based missile, the advantage is reversed. This probably remains true even though the initial cost of the ship can be rapidly depreciated with each missile launched. If, however, the vulnerability factor is added to the equation, the scales may be tipped back again. If fixed ICBM sites are to be wiped out in short order in a thermonuclear missile attack and if sea-based launchers survive destruction for any considerable time, the cost of the two systems per missile launched and target destroyed may well throw the fiscal advantage to the system with higher initial cost but lower vulnerability.

Submarine and surface ships are relatively slow, and their missiles are of only intermediate range, so that they will have to be positioned close to the enemy's coasts at all times if they are to strike swiftly and deeply at his vitals. Land-based ICBM's, on the contrary, being entirely self-contained weapons, can fly from one continent to another with incredible speed. They are always in position to fire, and if their sites survive a surprise attack, they can strike their retaliatory blows in a matter of minutes or hours. Unless sea-going missile launchers are in firing position when the surprise attack falls (and this seems unlikely for more than a fraction of the total force), it will be many hours or even days before their retaliatory missiles can reach the target. For the attacker, who has the advantage of being able to position his sea-going missiles in advance, ICBM's and IRBM 's can strike simultaneously; but for the victim of attack, the ICBM's have the advantage of instantaneous response, which the sea-going system cannot match. The whole matter seems, therefore, to turn on the survival factor. If the defender's ICBM's are sufficiently dispersed and in large enough number, and if they can be given any mobility at all so that the enemy cannot keep track of their exact coordinates, enough will survive to deter the enemy or insure some retaliatory capacity.

The combination of multiple missiles and nuclear power in the submarine has, nevertheless, created a weapons system of extraordinary potentiality. The capability of the missile submarine adds immeasurably to the ability of a naval fleet to attack distant inlands targets. There is little question that it is less dependent on base support and lines of communication than any weapons system in existence, and that it contains numerous advantages not available to its land-based counterparts. Yet, just as does any weapons system, it has its limitations. Like the bomber of the Strategic Air Command or the ICBM of the future, it is solely a retaliatory weapon, serving no useful purpose in other types of wars. Neither SAC nor Polaris-type submarines are dual-capacity weapons systems, and the latter, being restricted to use against stationary targets, lacks the versatility of manned aircraft. Given an arsenal of ICBM's which can long escape detection and destruction, our retaliatory capacity against fixed targets in total war will be flexible and potent, though the mobile bases of the enemy and the demands of limited war seem to require that some consideration be given to manned aircraft as well.

The question currently being raised about when and to what extent missiles will replace piloted aircraft must remain pure conjecture until the capabilities and limitations of the two are assessed with respect to the various missions now assigned to the manned plane. In general, missiles are capable of much greater speed and altitude and are less adversely affected by weather than are piloted aircraft. It also appears that, at least early in the period after ICBM's and IRBM 's become available, these will be less vulnerable to enemy countermeasures once they are launched. Indeed, missile bases, because of their small size and because of the ease with which they can be placed underground, should also be less vulnerable than air bases to enemy attack. Moreover, mobile missile launchers on land and sea have a greater capacity for escaping destruction than do fixed air bases.

Piloted aircraft, nevertheless, continue to have certain distinct attributes which missiles lack. The manned plane is not only much more accurate than is the long-range missile (an important characteristic in total and limited war alike), but it also carries a human crew which can respond to many situations that the electronic brain cannot master. With piloted aircraft, it is possible to halt a retaliatory attack initiated in error, and, in a world which lives under the necessity of hair-trigger and massive response to assault through the air, a system which permits the attack to break off may spell the difference between peace and destruction. Moreover, target selectivity is possible for manned aircraft: the pilot can shift his point of attack after the mission is launched to avoid wasting his payload on a target already destroyed or neutralized, and he can meet in other ways the needs of a rapidly changing strategic or tactical situation. Piloted aircraft are capable of carrying out a wide range of missions involving the need for human judgment or discreet control, such as reconnaissance, transport, tactical and logistical support, and damage assessment. Current offensive missiles, as well as those anticipated in the near future, lack these capabilities altogether. For the most part, they are vehicles designed to deliver highly destructive warheads to a fixed, selected enemy target and they are capable of playing no other role.

Given these characteristics, it is evident that military preparedness in the immediate future will require both types of weapons: 1.) the long-range missile group for use against large immobile targets where coordinates of launching and delivery points are accurately known; and 2.) manned aircraft for use against pinpoint and moving targets where extreme accuracy, maneuverability, and pilot judgment are essential. Both of these types will be necessary because: 1.) the ICBM's minimum margin of error of from two to five miles at a range of 5,000 miles makes it either uneconomical or ineffectual against all but large targets, except where saturation by radioactive fallout is the main objective; and 2.) the same applies to the IRBM on a reduced scale, for, though its accuracy is greater than that of the ICBM, it falls far short of the piloted bomber in this respect. The vulnerability of piloted aircraft to defensive missiles, however, and the invulnerability of offensive missiles to today's countermeasures, make it evident that the ballistic missile will be more effective than bombers against large, heavily defended, stationary targets. Against moving targets and small-scale fixed ones, on the other hand, the manned aircraft alone remains effective. The combination of piloted plane with air-to-surface missiles offers intriguing possibilities for incorporating the best features of both. For the foreseeable future, however, the need of both missiles and of piloted aircraft for missions now handled by the Strategic Air Command alone, seems probable.

For limited war situations and where careful target discrimination is demanded, long-range missiles have little application. On the other hand, the short-range missile family can add significantly to striking power and versatility in limited war, since these weapons possess the necessary accuracy and flexibility for such operations. But even in this role they will be employed primarily as supplements rather than as replacements for manned aircraft, while manned aircraft will continue to be a dominant factor in these important operations despite the fact that current surface-to-air missiles have already outstripped interceptor craft as short-range defensive weapons. For the fact is that fighters equipped with air to-air missiles are still superior to the defensive missile for use against enemy piloted aircraft, since their greater mobility and range give them the versatility to act under a variety of situations. Where, for example, defense in depth is necessary, the piloted interceptor will probably be required for some time to come.

For strategic air defense, manned aircraft also have the capacity to identify hostile planes at long range, thus averting unnecessary destruction of friendly craft without exposing defended areas to unopposed surprise attacks. Swiftness and length of range, plus accurate identification are imperative in the present world situation, and no solution to this problem would be readily available if we were to adopt missiles as our sole means of air defense. Against long-range missiles of the future, of course, the anti-missile missile will be the only sure solution. It appears, therefore, that, while missiles are certain to replace a substantial portion of piloted aircraft in air defense, the need for both in a closely coordinated system will continue for an indefinite period of time.

There is, indeed, some question whether missiles will ever completely replace the piloted plane, for capabilities of manned vehicles will undoubtedly grow, along with missile developments, much of the knowledge deriving from one being simultaneously applied to the other; and manned aircraft will continue to be advantageous in reconnaissance and for selecting targets en route.

Moreover, wherever human judgment and precise control are required, piloted aircraft will prevail if they can develop adequate defenses against the enemy's counter weapons. There is no theoretical reason why this will not be possible, for, despite the performance of missiles, breathtaking by current standards, their inflexibility lends credence to the prediction of effective missile defense. As defenses against missiles are perfected, it is conceivable that the pendulum may swing away from missiles toward renewed reliance on piloted aircraft in military forces of the future. Certainly, we cannot afford to confine our planning to a single concept, no matter how attractive it may seem at the moment. Thus, the continued development of both types of weapons, and their integration within our military forces, remains essential for national security in the future.

The fact that manned aircraft still have a vital role to play in both offensive and defensive strategic concepts lends considerable credence to the plaudits that aircraft carriers continue to receive as instruments of war. The nuclear-powered aircraft carrier, constantly on the move and having both intermediate-range missiles and manned aircraft aboard, will have the twin capacity of mobility and of great striking power so essential for the total war of the future. Moreover, the nuclear-powered aircraft carrier's capacity to serve double duty in limited and total war alike is of immeasurable advantage in a period of steadily mounting defense expenditures. The cost of this vastly expensive weapons complex can be applied to both types of military preparedness, and thus be relatively low for each. These are potential advantages which make it imperative for us to give close scrutiny to the claims being made for additional carrier appropriations.

The carrier is simultaneously a mobile air base (a unique characteristic in itself) and a mobile sea-based missile-launching site. The latter is an attribute the carrier shares with the Polaris-type submarine. The carrier's piloted aircraft, coupled with its own mobility, gives a precision and selectivity for waging limited war anywhere in the world. In this, it is certainly unmatched by long range missiles, and in certain respects it is also unequaled by land-based aircraft.

So much can be said with assurance. But additional claims by naval enthusiasts warrant closer examination. Carrier aircraft, it is said, have already put the Navy squarely in the strategic bombing business, and, with the addition of guided missiles of the Regulus type and of ballistic missiles like Polaris, the carrier of the future will have a total war offensive capacity of massive proportions. It would be more accurate to say merely that carriers have a considerable potential in this direction. In the first place, the enemy's short-range defensive missiles will reduce the present carrier planes' striking power to a marked degree. In the second place, carriers will have sufficient naval missions to keep their complements well occupied on tactical operations of both defensive and offensive nature. Finally, until restrictions are lifted from service roles and from missions which forbid the Navy from playing a major part in strategic warfare, carriers will never receive enough appropriations to fulfill the expectations of their proponents. Nevertheless, this mobile weapons complex, by its capacity to escape destruction from the long-range ballistic missile, could add both flexibility and weight to our retaliatory capacity, if it were assigned a major retaliatory role and given the funds to go with it.

The major question in regard to carriers is, of course, their defensive capabilities. It is claimed that the aircraft carrier, by reason of its special characteristics, has actually benefited far more from missiles than it has suffered from them. Not only is the carrier invulnerable to the fixed-trajectory, long-range missile as long as it stays out of port (which the nuclear-powered ship could accomplish for indefinite periods of time), but its own short-range defense missiles give it considerable protection from attacking aircraft. Such surface-to-air missiles as Terrier and Talos, which are currently being installed in carriers and support ships, along with Sidewinder, Sparrow, and other air-to-air missiles employed by modern, high-performance, interceptor aircraft, will sharply increase the effectiveness of carrier air defense against piloted aircraft. In other words, the carrier need not worry about the ballistic missiles of the enemy, and, if the carrier task force takes full advantage of the new defensive missiles coming into existence, it may also expect to decrease its vulnerability to attacking aircraft. With the IRBM to give it more deadly striking power and with short-range missiles to make it more defensible, the carrier is pictured offensively and defensively as the finest weapons complex of the missile age.

Under ideal conditions, there is probably a good deal of validity in this assertion. In a period when defensive missiles outmatch offensive aircraft, the roving carrier task force has a good chance to escape immediate destruction. The great military value of the carrier, however, combined with its slow rate of speed, will make it a tempting target for manned aircraft and submarines. Under sustained and concentrated attack by Soviet air and subsurface forces, operating with all the numerical strength they can command, carriers will not be unsinkable, nor does anyone make such an absurd claim. Rather, the argument is that the more enemy striking forces are drawn to our ships at sea and away from our cities and land-based launching sites, the greater will be our chance to survive and retaliate. This argument perhaps has some validity. The enemy's ICBM's and its missile submarines will not be diverted from our shores, but they may be more easily located and identified for what they are, if substantial numbers of its manned aircraft and conventional submarines are engaged elsewhere against our ships at sea.

It is doubtful, however, whether even the expert can speak with certainty on this point. Carrier task forces are immensely expensive to create and maintain. If destroyed with their ships, planes, missiles, and personnel, a major portion of the defense investment of the country goes with them. Whether their tremendous advantages in total and limited war and their high capacities for defense and offense warrant such an expenditure, is a decision requiring expert knowledge, strategic acumen, and absolute objectivity. The suggestion that they may act as decoys to save our cities should carry no weight, unless there is reasonable assurance that this would in fact be the case.

It is argued by some naval enthusiasts that the decision should go in favor of the carrier and Polaris-firing submarine. This combination alone, it is said, will be sufficient to maintain the peace by insuring the destruction of the enemy's fixed launching sites, while leaving our sea-based forces relatively untouched to strike again and again. The Soviet Union will not, however, be without mobile undersea bases of its own, and the launching pads of its ICBM's will be difficult to locate in advance. Moreover, the Soviets have a high capacity for waging limited wars. The Polaris submarine has the means to counter none of these advantages, and the aircraft carrier is probably too vulnerable to air and undersea attack to depend on as the sole instrument of deterrence or victory in an all-out war. To locate and combat Soviet submarines and land-based missile sites, as well as to fight limited wars, it appears that manned aircraft with protected bases remain essential.

To package in one instrument the capacity to locate unknown or moving targets with the ability to be free of base restrictions is the ideal which Americans seek in weaponry, and the proposal which is being offered to effect this wizardry is the nuclear powered seaplane. The principal reason advanced for selecting the seaplane rather than land-based aircraft is that even if the latter could remain aloft to the limit of human endurance it would eventually have to return to a base which could be obliterated, whereas seaplanes need no bases. They could be in a constant state of dispersion and could shift at will, thus providing maximum assurance that a surprise attack would not destroy our retaliatory power. It is confidently predicted that relatively few such aircraft, constantly on the move with their own transportable docks, establishing bases for themselves wherever a suitable stretch of water was available, would keep our counterattacking capacity high. The docks have already been developed; and if the nuclear-powered seaplane can also be constructed, a few of them, it is claimed, would give us a measure of guaranteed retaliatory power unmatched by dozens of vulnerable land-based craft.

Some people are already thinking about how to fit such seaplanes into our retaliatory system. They are proceeding on the assumption that the firepower of the present and the mobility of the future will cause the counterattack, or defense through offensive retaliation, to become strategically invalid, if strategists and policymakers continue to be hypnotized by the pseudo-mobility of a completely land-based air force. This line of thinking is all to the good, but the production of the nuclear-powered seaplane must wait upon the scientists and engineers and upon the cost accountants of national security. The problem entailed in putting a nuclear power plant into an air frame is not yet solved, and the cost relative to vulnerability over the target may make such a weapons system prohibitively expensive. Air defense missiles may well rule out an instrument as costly as a nuclear-powered plane, whether or not it carries its own base with it.

Production of nuclear-powered seaplanes is too far in the future to require their inclusion in any assessment of weapons systems made for current strategic planning. The major systems demanding immediate decision are those having implications for total war at the beginning of the missile age, for it is between these that the most basic differences of opinion have arisen to plague the policy-makers and strategists of the United States. Of those now operational or soon to be available, three are land-based: the piloted aircraft, the ICBM, and the IRBM; and two are sea-based: the Polaris-firing submarine and the aircraft carrier. Although the warheads of these systems vary in yield, all are so lethal in effect that relative destructive capacity can be ignored.

In terms of vulnerability in flight, it is clear that the piloted aircraft is far more liable to destruction than ballistic missiles of any kind, and that of the latter those of shortest range are least vulnerable, because their short flight time allows little warning. Among the various types of missile launchers, the submarine is unquestionably the most secure, with the huge stationary air base at the other end of the spectrum. Land-based missile launching sites, being smaller in size than fixed air bases, will fare somewhat better, and if they can be given some mobility in the future, their present vulnerability will drop sharply. The aircraft carrier, although relatively immune to missiles, operates under the shadow of a very serious threat from submarines and aircraft. If the main characteristics of deterrent forces are invulnerability, high striking power, and the capacity to draw strikes from the homeland, there seems little question that, for the immediate future, sea-based forces have an edge over land-based aircraft and ballistic missiles. If the land-based missile can be given mobility, however, its speed of response to attack will eventually place it high among the deterrent forces in our armory. The manned aircraft will probably grow increasingly invalid as a retaliatory force, because only in terms of accuracy does it outweigh the missile. If deterrence and retaliation were our only concern, the main emphasis in preparedness would be to strengthen our sea-going forces and to give mobility to our land-based retaliatory weapons.

This, of course, is not the case; national security will not be maintained under any single strategic concept. The primary requirement is an ability to meet a wide range of threats and to take advantage of whatever opportunities the future may bring. The most flexible weapons system today is the carrier. Its mobility and its capacity to launch both missiles and aircraft combine to give it a role in total and limited wars alike. Unlike SAC, Polaris, and the ICBM, which are solely instruments of retaliation, and unlike the ground forces, which must carry the brunt of limited wars, the carrier is capable of contributing to each kind of war. It is, however, too slow, too vulnerable, and too occupied with naval targets to be our sole reliance as a retaliatory force and is only one of many instruments necessary for limited conflicts. Yet, the additional strength it has derived from the development of missiles has given it a larger role than ever in total war, and a vital one in lesser conflicts.

The final factor which must be taken into account in making decisions on the development of these various weapons systems is the matter of relative costs, and this is perhaps the most difficult of all to assess correctly. One reason for this is the unpredictability of attrition rates. Although a general order of relative vulnerability may be determined, this is scarcely precise enough to use as a reliable basis for cost estimates. A second reason is that some of these weapons systems can be employed in several types of missions, making it difficult to prorate their expense to any single one. Development costs are also difficult to pin down to specific weapons systems, and so are the support costs they generate. In a period in which new weapons and counter weapons are being introduced with such incredible speed, it is, moreover, impossible to determine obsolescence rates, so that any calculation of depreciation against original cost can only be an educated guess at best. Finally, cost must be measured in terms of results achieved, and since this cannot always be determined before the weapon is put to use, often the best that can be done is to consider the cost of any weapons system to be acceptable if it is assumed to be essential for the nation's security. Beyond that, the emphasis to be placed upon the production of several essential systems must be determined by the relative effectiveness of each.

It is clear that neither carrier nor missile submarine, land-based plane nor ballistic missile, has a monopoly on effectiveness. All have sufficient merit to be included in the arsenals of the present and immediate future, and each has deficiencies which make the others essential for the maintenance of security. Moreover, the fact that they all differ in operation, and in the threat they pose, complicates the enemy's defense problems. This alone might justify the development and operational readiness of all of them, but the major reason why none should be discarded yet is the need to avoid the pitfall of rigidifying our strategic effort, of pointing it in a single direction, and thus making our security dependent upon a single set of circumstances. Were the United States to fall into this trap, and the enemy achieved a major breakthrough in offensive or defensive weapons, we should be at his mercy.

This is not to say that each system should receive equal emphasis either today or throughout time. The land-based strategic bomber is not yet obsolete, although it is already in a state of obsolescence and can be expected to receive a decreasing proportion of the defense budget in the years ahead. The long-range missile family and anti-missile missiles must be developed, improved, and placed in quantity production as rapidly as possible without regard to cost. This applies equally to the Polaris submarine, at least until it becomes certain that land-based ballistic missiles can be given highly mobile launching platforms. Even then the submarine's shorter-range missile launched from neutral waters will give the enemy special defense problems while freeing us from the political implications of depending upon bases overseas.

The future of the carrier task force is less predictable. At present, it is a weapons complex of tremendous versatility and striking power, but the early years of the missile age, which have given it additional significance and strength for total and limited war, will give way to time. No one can now predict what future weapons developments will do to its capacity to survive attack. All that can be said today is that, being a mobile base for both plane and missile, it has alternate means of insuring its offensive characteristics, while its future defensive capacity is shrouded in uncertainty. Until that potential weakness becomes substantiated in fact, however, it deserves a share of the military appropriations allotted to the strategic plans for both total and limited wars.

The purely military implications of guided missiles, as the foregoing has suggested, have attracted by far the greatest amount of attention in the United States. This, of course, is not surprising. The military significance of the new rocket weapons is such that the sorely beleaguered strategist now finds himself confronted with problems of virtually insuperable complexity. However, it seems certain that the missile age is ultimately going to produce some of its greatest challenges in the political area, both national and international. Thus far, it must be admitted, the general political effect of the guided missile has been, as with its military implications, to intensify familiar problems and existing issues, but one can already detect certain political areas in which rocket weapons are beginning to produce radical change.

American rockets, for example, have started to score some impressive hits upon purely political targets, and the guided missile has already proved to possess high potential for partisan political warfare. Arguments about the various guided missile programs have intensified the frequently bitter debates of recent years about the state of American military preparedness. Prospective Presidential candidates add charges and countercharges about the missile programs to their statements and speeches about the defense effort. Likewise, one of the first effects of the guided missile has been to increase the familiar, long-standing competition between the Army, the Navy, and the Air Force to add, as it were, another dimension to a rivalry which has permanent existence regardless of the current organizational charts in the Department of Defense. It did not long escape military men in the Pentagon that, in their unending pursuit of shares of a limited defense budget, the service with the most impressive array of rockets would be in a strategic position to press its financial claims. The Army, convinced that it has come off third best in recent years, counted heavily on its missile program to regain lost stature and was understandably vehement when Secretary of Defense Wilson decided that the ground forces had no logical reason to control weapons with a range of more than 200 miles. And, as Professor Hammond's chapter indicates, the orbiting of Sputnik and the clear signs of Soviet achievement in rocket weapons provided the rationale for the latest in a continuing series of moves designed to create a more perfect unification of the armed services. Thus, the advent of the rocket has already created a "missile controversy," with implications not only for domestic American politics but also for the continuing rivalries between the services, and for the drive toward further unification.

A more significant problem arises from the way in which guided missiles create a further compression of the time factor in modern war and international relations. Time, of course, used to be the silent ally of the United States. Even in World War II, it was still possible for the United States to be an indecisive bystander for more than two years and yet ultimately to mobilize its resources in sufficient time to crush both Germany and Japan. There was still time to construct effective instruments for joint and combined operations after the United States had become a belligerent; the Joint Chiefs of Staff, the Combined Chiefs of Staff, and SHAEF were all laboriously hammered out after the "shooting war" had started. And there was still time for President Roosevelt to debate the niceties of target selection and strategic objectives with Winston Churchill.

Since World War II and, particularly, since the advent of supersonic jet aircraft, the time factor has been reduced almost to the vanishing point. It was evident a decade ago that, in the event of another major war, there would simply not be sufficient time to form a coalition or to construct combined military institutions after hositilities began. Indeed, one of the strong arguments in favor of NATO is independent of its forces in being; much of NATO's value arises from the fact that an interallied command exists, that experiments in international military cooperation have been carried out, and that experience has been gained. Likewise, the importance of time plus the increased destructive power of nuclear weapons has long since led the United States to abandon the idea of a ponderous national mobilization, and to base its security instead upon the concept of both instant and massive retaliation.

The guided missile merely shrinks the time factor from a matter of hours to one of minutes. It therefore does not revolutionize the dimension of time in international affairs. But this last shrinkage points up some existing political problems and may even compel American leaders to face up directly to some issues which have hitherto been compromised. Consider, for example, the question of the control of atomic weapons. As long as there seemed to be a few hours in which to make decisions, it appeared feasible for the United States and its NATO partners to live with some rather makeshift arrangements. Since Congress was reluctant to share nuclear weapons, atomic warheads were neatly stored at foreign bases and kept under American control. At the moment of ultimate decision, presumably, the American guardians would hand them over to their European partners. This system, considering the speed of jet aircraft, was probably never really practicable; but, with the time for decision reduced to minutes by the guided missile, it should become obvious that no such system of control is militarily feasible. The rocket may therefore compel Congress to reconsider a problem that heretofore has not been properly faced.

The transition from jet bombers to guided missiles intensifies the problem of how and by whom the ultimate decisions will be made. With the jet, after all, it is still possible to think that there is, however small, a margin for error; even the manned aircraft of the Strategic Air Command can be recalled to their bases in the event that the wiggles on the radar screen do not turn out to be Soviet planes. But the decision to launch a guided missile is irrevocable. Once in the air it cannot be called back. The rocket, in short, starkly underlines the fact that political decisions must be made in advance, must be instantaneous, if not automatic that there is no time for allies to consult with one another, no time for legislative bodies to deliberate, no time for the man in the street to decide if his interests are affected. And, it might be added, no time for the niceties of target selection. If a future British prime minister thinks that Bulgaria is a better target than the area around Leningrad, he will not have the opportunity to press his point, at least not until after the first mushroom clouds have cleared when it may make no difference.

In any event, this final compression of time raises a host of political incalculables. Given the need for automatic decisions about the kind of enemy action which will trigger a response by guided missiles, the question arises, for example, whether or not the European countries will willingly grant such power to the United States. Or, when the European countries themselves possess rockets armed with nuclear weapons, how will the United States adapt to a situation in which its allies have the power to make the ultimate decision? These issues, incompletely faced in the jet age, will press with increasing urgency in the missile age.

The rocket likewise intensifies problems in planning the defense of the Western Hemisphere, and, like the supersonic bomber, it has already affected relations between the United States and Canada. Missiles and jet aircraft are no respecters of the artificial boundaries which exist between countries; launched across the polar ice cap, they will be sublimely indifferent to the "unguarded frontier" between the United States and Canada. Any defense against Soviet attack must be a North American defense, not a Canadian or an American defense. The impact of the intercontinental bomber has already led to the establishment of a joint Canadian-American air defense command NORAD, patterned on the familiar model of SHAEF and NATO in which defense sectors, interception points, and radar controls are established as if no political frontiers exist. The advent of the missile simply confirms the necessity for this arrangement and makes it irrevocable. Both Ottawa and Washington have agreed to share the responsibility for air defense; in the future, they cannot afford basic political disagreements or different political strategies. The guided missile, in short, completes a development which should have become obvious with the intercontinental bomber; it removes one of the principal elements of national sovereignty the ability of a single nation, acting on its own initiative, to determine the question of peace or war.

The guided missile has already begun to have a marked effect upon international politics. Indeed, it has already affected at least the edges of American policy in one international crisis of recent years. Many considerations led the United States, in the autumn of 1956, to compel the British and French to abandon their ill fated attempt to discipline Colonel Nasser; one of these was undoubtedly the Soviet notes to London and Paris which contained thinly disguised hints at Russian rocket retaliation. The receipt of these notes, it has been reported, caused the greatest consternation in Western capitals since the outbreak of the Korean War and, without question, added to the American desire to terminate British and French military action. Russia's willingness to use guided missiles to engage in "atomic blackmail" was a new and foreboding tactic in international politics.

One of the crucial questions for both political and military leaders has always been the location and availability of overseas bases as points from which a nation's military power can be brought effectively to bear upon an enemy. As far back as the early years of this century, the United States Navy found its dream of exercising real authority in the Far East vanishing because America possessed no base nearer China than the Philippine Islands. Since World War II, the very bedrock of American security has been the series of air bases which ring the Soviet Union and without which the Russians might possess retaliatory superiority.

The existence of these bases has, over the last decade, produced political tension and friction and, in many countries, has created considerable fear in the civilian population. All of these problems are greatly intensified when it is suggested that the United States would now like to establish a missile base in some foreign country. Theoretically, there appears to be little difference between permitting the United States to build a base for guided missiles and allowing this country to construct a base for manned aircraft. Indeed, some American officers have already expressed polite surprise at the reluctance of some Western European countries to cede such sites. Faulty diplomacy by the Department of State, a not unusual charge in these days, has been blamed; and the ingenious suggestion has even been made that, since rockets require less launching room than manned aircraft, the Europeans should welcome missile sites as a way of reclaiming scarce agricultural land.

The real problem is the attitude of civilian populations and what the average European thinks the consequences will be. To many a European the guided missile armed with an atomic warhead - raises the moral issue of war in a form more acute than does any other weapons system. But the crucial fact is that the rocket finally brings home what should have been clear when the first American bases for manned aircraft were built in his country: the fact that, in another major war, the first move of the enemy will be to destroy the bases from which any retaliation can be launched. Moreover, the rumored inaccuracy of the guided missile makes many fear that the areas of destruction will be far more extensive than if an attack were delivered by conventional aircraft.

The advent of the guided missile particularly, the development of rocketry in the Soviet Union has already begun to undermine the political value of our own deterrents. As long as the United States possessed an atomic monopoly, and, indeed, as long as America possessed a clear superiority in delivery systems and retaliatory power, it was possible for Europeans to convince themselves that a third world war might conceivably take place without wholesale destruction of the Continent. But the recent evolution of the missile race has destroyed the last lingering traces of such illusions. It is clear, for example, that, for some time to come, there will be a period between the obsolescence of the Strategic Air Command and the development of the ICBM when American retaliatory power will be heavily dependent upon the inter mediate range missile based in Europe. And it is also clear that the Soviet Union possesses the capacity to retaliate effectively against these intermediate-range rocket sites. Europeans are therefore giving far more serious thought to the implications of missile sites than they did to the issue of bases for manned aircraft. Moreover, the evidence of Soviet missile capacity tended to make Europeans even more fearful of American diplomacy and of the late Mr. Dulles' penchant for operations on the brink. As indicated by the most recent crisis over Quemoy and Matsu in the fall of 1958, many Europeans simply wanted to look for storm shelters and to find almost any solution short of war. The credibility of the American deterrent is also being more frequently challenged. Europeans increasingly wonder if American power, as presently constituted, will actually deter the Soviets and, indeed, if it will ever be employed as heretofore announced. The guided missile has, therefore, not only intensified the problems of overseas bases, but has also started to undercut the basic American position in Europe.

It is not surprising that the guided missile has already added a new dimension to the various proposals to relax East-West tensions by "disengagement," i.e., by creating a neutral zone in Central Europe. George Kennan's well-publicized arguments for disengagement were founded to a considerable degree upon the argument that the establishment of missile launching sites in West Germany would only lead the Soviet Union to build comparable bases on the other side of the iron curtain. As a result, Kennan maintained, an even tighter lid would be clamped upon the Soviet satellites, the chances of war by miscalculation would be heightened, and the prospects of an eventual German settlement would be virtually destroyed. The United States, Kennan continued, should therefore abandon its plans to obtain missile sites in Central Europe. But proposals for disengagement have encountered heavy opposition. Many military leaders apparently fear that, in a period when the United States must depend heavily upon inter mediate range missiles, the loss of rocket launching sites would cripple American retaliatory power. As Kennan himself has written, there are many who see "any modification of present NATO military policies as possible only when the United States can be sure of matching the Soviet Union in the field of intercontinental ballistic missiles."

Yet, advocates of disengagement as well as those concerned about the political effects of air and missile bases in Europe count heavily upon the next stage in missile development to aid their point of view. Kennan, for example, writes hopefully of the ICBM and the Polaris system, whose development, he reasons, will lessen the importance of European bases and make it possible for statesmen to begin considering a neutralized Central Europe with less trepidation. And it has also been argued that greater reliance upon long-range and sea-based missiles would free the United States from many current and embarrassing political commitments from the unhappy situation in which Mr. Dulles had to defend Portuguese possession of Goa because of American dependence on bases in the Azores. Indeed, there are already a few Americans who are hopefully looking forward to some future date when the ICBM, launched perhaps in Utah but able to strike the Ukraine, will enable the United States to cut back its worldwide commitments and return to the concept of Fortress America. To this small group, the great advantage of the ICBM is that it will thereby make at least a modest amount of isolationism once again seem respectable.

It is not possible at this time to write with any degree of assurance about the ultimate political implications of the guided missile. In this early stage of the "missile age," the political overtones are far from clear. It seems fantastic, for example, even to suggest that the ICBM might serve as a means for even a modest return to traditional American isolationism. Yet, one can at least imagine a future situation in which the United States and its European allies are in disagreement, in which Soviet missile strength makes the defense of Western Europe appear virtually hopeless, and in which the tax burden on the American public has steadily increased. In such circumstances, it is not wholly inconceivable that the advocates of Fortress America will begin to find increasing audiences. Less speculative than this, however, is the fact that the present American dependence on the intermediate-range missile means that we may actually find ourselves more, rather than less, dependent upon the good will of our allies. Indeed, whereas in the early days of the Eisenhower Administration there was considerable talk about an "agonizing reappraisal" of American commitments in Europe, the NATO meeting of late 1957 found this country actively seeking European approval for missile bases. Whether the political result will be for good or evil as yet remains to be seen. Some observers, like Kennan, think that the intermediate-range missile will exacerbate European political problems and make it impossible to ease tensions between East and West. Others have argued that the political bargaining power of the European states will increase and that the Europeans, for the first time in over a decade, will have a check upon American policy. And this, if nothing else, might add a certain amount of healthy flexibility to an international situation so long dominated by two superpowers.

One of the great dangers of the "missile age" is that in the desperate race for rocket superiority, and in the frantic attention being paid to the military significance of these new weapons, their political implications may be submerged, and military factors will take precedence over the political in our minds. Guided missiles are, indeed, so new and so potentially deadly in effect that they carry with them a kind of repulsive fascination a fascination which diverts us from their more somber and less dramatic political consequences. The very deadliness of their military characteristics, nevertheless, makes the maintenance of something less than total war imperative, and gives added meaning to the old axiom that military considerations must be kept subordinate to political considerations in the minds of strategists and statesmen. This is more true today than ever before; it is also more difficult to keep in mind, and less easily put into practice, at a time when weapons are so lethal and international tensions are so acute.

8

Influence of Modern Weapons on Security

Introduction

The strategist must attain a clear idea of the direction in which the tools of war are leading him if he is to avoid responsibility for a world destroyed. The weapons of the modern era have already had certain tangible and baneful effects upon strategic planning. They are having, and will continue to have, a profound impact upon strategic theory, upon the whole process of thinking out just what war is and what strategy means in the nuclear age. It is important that these effects, both the practical and theoretical, the sinister and the salutary, be spelled out, for Americans by and large have failed to consider them sufficiently, and, as a result, are overlooking their far-reaching implications.

It is evident from previous chapters in this book that traditional American conceptions of peace and war are having a strong and sometimes injurious effect upon current strategic planning in this country. The adaptation of nuclear power to the field of war, far from shattering these conceptions, has reinforced them. Thus, one of the strongest influences that weapons are exerting upon strategy is being effected by this correlation. Americans, being defense-minded, have been led of necessity to the strategy of instantaneous and massive retaliation by the destructive capacity of the weapons now in the hands of the enemy. This has never been an entirely satisfactory solution to the major problem of the nuclear age, and Americans are just now beginning to seek a happier one in the development of a weapons system so mobile that it can escape destruction. Moreover, since they consider peace and war to be completely distinct entities totally divorced from one another by the gap between good and evil, Americans have been brought by their nuclear weapons to a strategy of all or nothing, deterrence or defeat, capitulation or suicide. Yet, every day it is becoming more evident that this narrow range of alternatives is unacceptable. As a result, we are being compelled by the very nature of weapons development to reshape our thinking about war itself in order to allow ourselves greater freedom of strategic maneuver.

Furthermore, while these new weapons have, in the last ten years, made us think of strategy primarily as a science of destruction, they are now beginning to make us revert to historical experience and to think of it once again as the art of control. In order to effect this transition, of course, our present tendency of permitting weapons to dominate and determine military strategy will have to be reversed, and military strategy in its turn will have to be restored to its proper place as only one element of national strategy. Finally, modern weapons are making necessary a renewed appreciation of the role of logistics in strategy a role which Americans have tended to forget or to submerge in their fascination with nuclear power as a destructive force rather than as a means of propulsion.

It should be evident even from this brief summary that the effect of the new weapons has thus far been to strengthen certain American misconceptions about the dual nature of war and peace, and thus to make us think about strategy in terms of black and white, when, by all rights of logic, it should be realized that strategy is every day becoming more complex, less a science of measuring degrees of destruction, and more the art of maneuver and control. The fact is that the revolution effected by weapons has been so swift that it has been difficult to keep pace with events. It has tended to make us disregard the lessons of the past and to consider the weapon as the primary gauge of strategic theory. In short, it might be said that intellectual insight is the missing element in strategy today. It seems equally evident, however, given time for reflection upon the true nature of strategy and on our own peculiar strategic problems, that weapons can be developed and strategic theories formulated which will help us to escape from the most dangerous dilemmas of the nuclear age.

The current doctrine of retaliation is a case in point. American military men, knowing that they cannot strike first because Americans will not condone aggression in any form, have felt compelled to consider above all else the offensive and defensive implications of the new weapons of mass destruction. They know from history that increases in firepower have generally favored the defense, as was the case in World War I when artillery was lined up hub to hub, and human flesh could make no headway against it. They know, on the other hand, that advances in mobility have generally favored the offense. During World War II, when aircraft flew over the lines and when tanks sliced through them, the offensive was rehabilitated even in the face of increased firepower. And in the postwar years long-range bombers with nuclear payloads have made the offensive vastly superior to the defensive. Again, historically, when the offense has outstripped the defense, the classic strategic solution has been to meet the aggressor with a counterthrust. This fact has naturally led American strategists, with their instruments of great firepower and mobility, to the doctrine of retaliation, for if the United States would not attack first and could not escape untold destruction from the onslaught of enemy hydrogen bombs and bombers, then the only course of action seemed to be to make its own offensive capacity so swift and devastating as to deter the Soviet Union by insuring its defeat along with our own.

This is what modern weapons have done to strategy thus far, but the situation has scarcely been thought through completely, nor has due weight been given to the weapons systems of the future. The current strategy of the United States, depending as it does primarily upon a large number of fixed air bases spread throughout the free world, may be sufficient today, but it will not be adequate for the future. In this second decade of the nuclear age, firepower has already reached a capacity for destructiveness which is at the point of diminishing returns, but the world is just on the threshold of the absolute in range and speed. When missile speeds become a commonplace among the great powers of the world, and when intercontinental ballistic missiles with nuclear warheads are capable of being laid on targets in advance, fixed bases will scarcely remain retaliatory assets. They will be easy to locate, difficult to defend, and vulnerable to a single hit.

The ICBM of the future will provide the enemy with an opportunity to destroy American retaliatory power before it is fairly launched. With the degree of speed, range, and accuracy forecast for these weapons of the future, the defender's ability to protect himself may well decline to zero. Under such circumstances, the United States would have to strike first or capitulate. Either of these courses of action would be acts of desperation to be avoided at all costs. The aggressor, by incurring the onus of the entire world, would lose the peace by the very process he employed in winning the war.

There appears to be only one way out of this dilemma, unless all nuclear powers reach such a harmonious relationship and a condition of such mutual trust as to permit disarmament. The alternative is to give our weapons complex so much mobility that it will guarantee our ability to retaliate under any circumstances; that is, the launching sites from which our planes and missiles are sped upon their way must be made sufficiently mobile so that they cannot be quickly located by the enemy and destroyed. The ability to move our retaliatory weapons systems about at will is the great imperative of future planning, for speed, range, and firepower will count for little if base mobility is lacking.

A number of proposals are currently being advanced which, their advocates believe, provide hopeful solutions to the problem of base mobility, and all in one form or another make use of the sea as the medium of maneuver. They include the aircraft carrier, the missile-firing submarine, and the seaplane, each given unlimited range by means of nuclear propulsion as well as relative invulnerability to missiles by the ease with which they can roam the great ocean areas to escape detection and destruction. Whatever merits these weapons systems may possess will be most pronounced in the missile age the age to come for, if they are to be worth the cost of development and construction, it will be primarily as counters to ballistic missiles in the hands of the enemy. They are, therefore, most appropriately discussed in connection with the strategic implications of missiles a subject of such complexity and importance in itself as to merit separate treatment elsewhere.

It is sufficient here to insist merely that the final decisions on the production of such weapons, as indeed on all future weapons, should be based on sound strategic calculations regardless of the contributions they happen to make to sea or air or land warfare. Strategy, it has been truly said, is indivisible; it should be considered as a unit rather than as a cluster of separate elements over which each military service attempts to exercise exclusive jurisdiction and control. Under an integrated system of strategic thought, policy-makers will not concern themselves with the advantages of a particular weapons system and with whether it happens to be land-based or sea-borne or is an aspect of air power; instead, they will consider what strategy must accomplish in a political sense, and, having taken due account of what means are available to achieve objectives, will then press for the development of whatever weapons complex that strategy dictates. The strategic bomber, whether land- or sea-based, and the medium- and long-range ballistic missile, whether launched from the ground, the carrier, or the submarine, are generally referred to as strategic weapons; they comprise the instruments of retaliation in total war, and should be considered solely in terms of what contributions they can make to that particular type of strategy.

The construction of weapons which will guarantee the ability to retaliate is, however, just a first step, not the last, for even if the enemy is deterred from launching an all-out attack through fear of committing suicide, there are other ways of waging war which the United States must be able to counter. Moreover, the enemy may not be deterred. If Americans should restrict themselves to a retaliatory strategy, they would be wagering their physical plant, their governmental structure, and indeed their whole way of life against the enemy's. To put the matter mildly, these would be unfavorable odds. Military experts know that to risk their fleets and armies unnecessarily is an unforgivable error, and they seek to accomplish their missions by means which will reduce the hazard. How much greater a blunder it would be for a nation to risk its existence by deliberately depending on a strategy which permits it no alternative but annihilation or capitulation in the face of every threat.

This nation has the highest standard of living in the world; it has a democratic philosophy and system of government which is perhaps the greatest tribute to man's ingenuity ever devised. For the United States to throw these assets on the bargaining table as equal stakes against the poverty and human degradation of the Soviet system would be the height of folly unless its vital interests were imperiled. Indeed they will not be risked if the Russians act only in areas where those interests are of minor import. The principle of deterrence is, after all, based upon the will to act. If the United States is unwilling to set its rigid strategy in motion except where its vital interests are threatened, it will lose its minor assets by default. It is worth noting, moreover, that a series of minor losses can have important cumulative effects and may even culminate in a vital loss to the free world.

It has been said that the Strategic Air Command is an "insurance policy" protecting the United States from nuclear weapons in the hands of others; it has been called a "police force" to promote world stability. Both terms are semantically incorrect. SAC does not insure the country against assault, nor does it pay off in future security and stability in the event of attack; it is a desperation strategy, pure and simple. The purpose of a police force, on the other hand, is to keep order by preventing crime and apprehending criminals, but strategic air power cannot possibly play all these roles. It can perhaps prevent major crimes and punish them, but chaos, not world order, would result from nuclear war. Strategic bombing is a method of employing weapons in one particular type of war. The very existence of SAC may channel future wars into something less than total conflicts, and this is its great asset, but more should not be claimed for it than it can do, thus permitting the nation's strategy to be rigidified. The United States must still have the weapons and the strategy to cope with the lesser and more subtle conflicts that may arise.

The process by which Americans arrived at this all-or-nothing thinking is as clear in origin as it is forbidding in result. It began with an inherent belief in the dichotomy between war and peace, and was then strongly reinforced by the awesome spectacle of nuclear energy in weapon form. From the moment the first atomic bomb was dropped on Japan, Americans became hypnotized by the power they had created. Having a monopoly of this power they became complacent. Even when the threat from the Soviet Union began to cast its shadow across the free world, there was only a momentary twinge of uneasiness. Korea demonstrated that a nuclear strategy was unsuited to all contingencies, and the nation was forced to build up its conventional forces in the Far East and in Europe, but this did not last. Gradually, it dropped back to the sterile concept that all but the most powerful weapons were futile, until today the nation's newspapers and military journals are predicting that the eventual use of atomic weapons in war is inevitable. Military leaders, government officials, and news analysts are saying either in criticism or accord: "A major war must now be atomic in nature. Why? Because the posture for atomic war cannot be reconciled with retaining a conventional capability." They are saying, in other words, that, since we have nuclear weapons, we must use them, because they impose on us a nuclear strategy which leaves us no alternative.

Korea, Indo-China, Suez, Algeria, and Lebanon are vivid demonstrations that a strategy of all or nothing, of peace or total war, is unimaginative, unnecessary, and inappropriate to the nuclear age. So blinded are some of us, however, by the brilliant glare and the mushroom cloud of the big bomb and its influence on strategy that we cannot see what is staring us in the face. Even today, when our monopoly of nuclear power has been broken, we maintain faith in a strategy which had only limited validity when the United States alone held that power. Dazzled by our nuclear weapons, we continue to base strategy too much on what these weapons and no others can permit us to do. Nuclear weapons, then, have become an end in themselves rather than one of many means to attain political ends.

This hypnosis exercised by weapons of mass destruction, by making military strategy too dependent on one set of tools, has deprived the nation of flexibility. With a rigid military strategy, no policies can be formulated and no objectives can be achieved which cannot be implemented by nuclear destruction. It is one thing to negotiate through strength; it is quite another to negotiate on the basis of a power which breeds self-destruction. Americans today are not only power-minded but nuclear power-minded. Their rigid strategy did not help them in Korea; it gave them little if any additional strength to deal with the Suez situation, and none for the

Hungarian crisis. In a sense, it has tied their hands by tying their minds to a single objective: total peace or total war deterrence or self-destruction.

This is what the development of nuclear weapons has done to strategy to date: it has rigidified it and left in its wake a policy of deter or die. Such a situation is obviously unacceptable, and strategists, therefore, are being forced to reshape their thinking not only about strategy but, first of all, about war itself. Their task is to find ways to permit greater freedom of maneuver by providing more alternative objectives. There is nothing inherent in the weapons system, in strategy or policy, or in the international situation which automatically restricts us to the extreme alternatives of peace or total war. Nothing limits us, that is, but our range of thinking and an unwillingness to alter our thought processes.

The expression, "there is nothing either good or bad, but thinking makes it so," has application here. If the statesmen and soldiers of the world are convinced that total war is inevitable, it will be; they will make it so by adopting strategic plans which permit no alternative.

Had there been no Korean conflict, the idea that war in the nuclear age must be total might have proved impossible to shake, so deeply imbedded was it in the national psyche; but there has been a limited war waged by a nation possessing atomic power, and more recently there has been one in the Middle East. These two cases alone provide ample demonstration that even with nuclear weapons wars do not have to be carried to the extreme. They can be, and they may be; but they do not have to be.

If the chance of nuclear war is to be minimized, alternatives will have to be provided, and thinking must start not with wars as ends in themselves but with the objectives which they serve. Since war as a rational act is intended to maintain or strengthen national security, nuclear conflict can be employed only as a last resort, because the most it can do is to preserve the sovereignty of the victor. It will virtually destroy everything else. Certainly, nuclear war cannot leave a belligerent better off than before the war; it can only leave him slightly less devastated than the enemy. The meaning of the phrase "to win a war" has, then, been sharply curtailed. The more limited war, the more advantages victory can bring, because the winner of a limited conflict can gain in almost every sense of the term, while the victor of a total war cannot.

The problem can be thought through clearly without falling into the line of thought which once caused a German general to say: "The best that diplomacy can do is to create the most favorable situation for military action." To be sure, this is one objective of diplomacy to build alliances and establish situations of strength, but the primary objective of both diplomacy and war is to foster circumstances which will permit a stable peace. Foreign policy must see to it that the United States does not become isolated in an enflamed world, for this would engender conditions for total war. Likewise, strategic plans must be so framed that there is something other than wreckage to work with when peace is restored.

In other words, there must be an end to the idea that policy governs only until war begins, that it is then discarded and war emerges as an end in itself. To prevent such a concept from dominating the minds of those who frame national security policies, the first thing that must happen is for them to stop thinking in terms of complete peace and stability on the one hand, and uncontrolled and unlimited war on the other. At a time when weapons have such a strong influence on strategy, and when the single act of target selection can destroy the world or prevent postwar stability, it is dangerous for the policy-maker and the strategist to think in terms of extreme alternatives, although the tactical commander may have to do so.

It is time to realize that there is a yawning gap in thinking between peace and unlimited war, that there are middle degrees of conflict, and that man can control his destinies only by keeping a firm hand on these lesser conflicts. He should be prepared to deal with realities and normalities rather than with abstractions and extremes. Complete peace and absolute war have never existed. Total absence of conflict and total absence of war to the point of suicide are meaningless abstractions. Human nature, with its appetites, jealousies, and passions, is unlikely ever to allow complete harmony to prevail. Nor, if mankind is to be credited with any degree of logic and intelligence at all, can we conclude that world suicide is a likely choice. It is the middle degrees of conflict between these extremes that should concern the policy-maker, for these are the real areas in which he has to operate.

The actual situation reveals a continuum of conflict, starting with a maximum of harmony at one end and graduating toward increasingly intensive and broadening strife across the spectrum. These degrees of conflict might be described in this way: first, there is the state of international tension at a level which still permits cooperation. Clashes in this range may be economic or political in nature; they can have to do with trade barriers or boundary disputes, and they may be local, regional, or global in scope. This is a state of affairs in which man still has control of his actions and emotions. He is still using logic and reason to arrest dispute or wipe out its cause. This, historically, is the most normal state of international affairs controlled conflict characterized by varying degrees of cooperation, compromise, and the will to reduce tensions.

The next state of conflict is one that the world has come to call "cold war," but that any previous age would have characterized in quite different terms. It has been given this label because two great power blocs exist so violently opposed to one another that, but for the presence of nuclear weapons, they would long ago have been engaged in direct military action. The name given to it is unimportant as long as we understand that it is a situation of tension so brittle and acute that it has been kept from getting out of hand with only limited success. It might well be called semi controlled conflict, for, despite an arms race and threats of retaliation, brushfires occasionally do break out. Persuasion and pressure rather than cooperation characterize this conflict stage mainly diplomatic persuasion and economic pressure, with frequent threats of military force.

The last stage in which some degree of control is exercised is that called "limited war." Coercion rather than persuasion dominates this phase. It is one in which the final vestige of control is easily lost: one false move, one heightening of objectives, one wrong weapon employed, and the war of extinction can come with startling speed. Coercion gives way to destruction, and even if the world survives, the chance for a stable international order has been shattered. In limited warfare, there is, however, still the possibility of restoring control, and there is everything to gain by so doing. This stage is man's last hope of survival.

Our efforts, then, and this includes our strategic thinking, should be maximized in the direction of keeping affairs at the lower end of the conflict spectrum. As long as we limit the intensity of conflict, we control our destinies, and strategy should be planned with this in mind. It is a fundamental error, and a dangerous one, to base strategic planning exclusively on that extreme state of affairs in which control has already been lost and destruction is inevitable.

Strategy should therefore not be considered as a science of destruction; rather, it should be thought of as the art of control. Nuclear weapons have, however, led Americans to think of strategy in terms of mathematics so many miles of range per gallon of fuel, so much speed per unit of energy, so much destruction per target selected and pound of explosive dropped, so much mechanical efficiency per dollar spent. The machine replaces the man; the manned aircraft gives way to the guided missile. There is less thought about maneuver, flanking movements, and double envelopments, and more about the straight punch.

Strategy has reached a low point indeed when it comes to depend on machines to destroy rather than on men to maneuver and control. It cannot fulfill its true function under such circumstances, for, by very definition, strategy is the manipulation of resources to achieve desired ends in situations of conflict. The element of conflict or competition is part and parcel of any strategic situation. This is why strategy is frequently referred to in connection with cards, sports, and politics. In such cases the attempt is made to outwit or outmaneuver the opponent. We play for victory if the resources, the power, the cards, or whatever are available, but, lacking these prerequisites, we adjust our strategy to cut our losses. Odds are figured, risks run, attempts made to exercise control over the action of our opponent by bottling up his resources, by hobbling his strength, by depriving him of initiative. The design is not to destroy him, but to deprive him of the ability or will to continue the play by destroying or curtailing his resources.

Thus it is with the strategy of a nation seeking to maintain its security in a world of strife. By means of its strategy, the nation attempts to exercise three kinds of control: first, it tries to direct or control its own resources, weapons, and power; second, it seeks to control or restrict the use of its enemy's resources by destroying them or making it unprofitable to use them; and, finally, it attempts to control the conflict situation to channel it into lines most profitable to itself. Never before in history has the element of control been so important, for nuclear weapons have made it imperative that the conflict be kept in the lower ranges of intensity. The objective of strategy, whether national or military, is so to control one's own power and that of the enemy that the conflict will not get out of hand and erupt into total war.

This can only be effected by keeping the national interest always in view, by realizing that the primary objective is to maintain our way of life, and that the only way to do this is to keep the struggle between the enemy and ourselves within reasonable bounds. A large stride has already been made in this direction once it is recognized that we are not in an all-or-nothing situation, but that the degrees of conflict are multiple and ever-shifting.

This is so because, in order to exercise control, there must be alternatives alternative courses of action, of means, and of goals. If the immediate objective is total victory and the only tools are hydrogen bombs, there is no alternative but total war. If, on the other hand, we set forth a goal which lies somewhere on the wide spectrum of conflict other than on its extreme destructive end, we immediately give ourselves room for maneuver. Courses of action then lie open in the economic, diplomatic, and psychological realms, with all the tools available to those fields of action. Moreover, a wide spectrum of alternatives exists in the military field, ranging from simple displays of power to limited, peripheral wars. In all these cases, a good military strategy retains a tremendous ability to maneuver and a capacity to keep the conflict under control.

The hydrogen bomb, by endowing the world with the means of self-destruction, is forcing the world to control its most devastating resource, and has thus reopened for the military strategist vast new demands for weapons development and unparalleled opportunities for exercising imagination in their employment. This should be an exciting and inspiring age for the strategist: never has he had a greater variety of tools in the field of communication, transportation, and weaponry; his theater of operations is larger and yet more accessible than ever before; his ability to act in the air, on the ground, and at sea has never been greater. It is the weapons of mass destruction which have forced this challenge on him, for never before has he had an enemy which can destroy his country in a matter of hours if he makes a simple mistake. His mental energies must, therefore, be directed as much toward controlling the enemy's actions as his own. This has always been true in war, but never to the extent it is today; strategy as the art of control must, therefore, once again come into its own.

Strategy is not, of course, simply a matter of using resources; it must also direct their development and their timely construction. No better illustration of this fact can be found in the postwar years than the summer of 1956 when Britain and France were envisaging military action in Suez. Neither country was ready at the time to apply military power of a type suited to the situation. Both could have employed strategic air power immediately, but this could not accomplish the objective of physically occupying the Canal and of keeping it open. What was needed was a combination of sea and airborne landings supported by all the paraphernalia of conventional warfare. Neither France nor Britain had the weapons, forces, or logistical capacity for such a rapid operation during the summer, and apparently they did not have it the following November when they tried to seize the whole Canal before it could be blocked. Their strategy was rendered useless and dangerous through lack of timely development of the means required to make it work.

The first thing, therefore, that the strategist must know is the kinds of results he will be expected to achieve. Knowing this, he can then determine the weapons and forces he will need. If he is to fight the kind of war which will create future conditions of peace, and if he is to maintain the flexibility necessary to meet a wide range of conflicts, he will need a wide variety of weapons. In order to be ready to fight instantly in any portion of the globe, under all conditions of weather and terrain, military hardware must be stockpiled in quality and quantity. If he is to strengthen the hands of the negotiators in cold war situations, the capacity of graduated deterrence must be his, for if the soldier can only react massively and not oppose lesser aggressions against minor interests, the hands of the diplomat will be tied. In other words, the strategist must maintain the capacity of employing selective force as well as massive retaliation, and both must be ready for instant use.

In order to deter all kinds of aggression and to keep the conflicts of the future limited, military leaders must, above all, have the means to make the enemy change his mind. In the nuclear age, their real task will be to destroy the enemy's will to continue fighting rather than his ability to do so, for, in order to destroy his ability completely, his cities or his military forces must be demolished; and with the Soviet Union as the enemy, this can evidently be done only by nuclear war. If, on the other hand, the enemy's will to fight can be curtailed, it will be unnecessary to destroy his ability and, in the process, destroy our own as well.

The means of curtailing an enemy's will to fight are numerous. The threat of massive retaliation is one important way; the ability to stop his minor aggressions in their tracks with precision weapons is another. Diplomacy, propaganda, and economic pressures may also be effective means as long as they are accompanied by military preparations which complement them rather than oppose them. The main thing to remember today is that, as soon as military destruction is substituted for more subtle forms of coercion and pressure, the ability to control the degree of conflict is lost, and so is the objective. In order to retain control i.e., in order to keep the war limit the strategist must have the capacity to fight both general and local wars; he must have ground, sea, and air weapons of infinite variety in order to protect his homeland and that of his allies.

Having developed the weapons needed to keep the conflict limited, the next step is to direct their employment in accord with the situation. Weapons themselves are inert until put to use, and one of the main tasks of strategy is to employ them correctly. Strategy, therefore, provides, among other things, the intelligent direction of weapons and forces for the achievement of a variety of goals. Again the element of control is emphasized, for strategy is an intelligent system of direction and control a means of convincing the enemy that it is in his interest to sue for peace. Yet, this will never be accomplished unless we first convince ourselves that the absolute destructiveness of total war is the negation of strategy rather than the furtherance of policy.

The nation which automatically jumps to employ weapons simply because they are stockpiled in its arsenals is substituting animal reaction for strategic thought. If it has failed to provide conventional weapons, it will have no choice; but with an adequate range of nuclear and non-nuclear tools of war, it has the opportunity for restraint. The weakness of the American armed forces in conventional categories at the opening of the Korean War made the decision not to employ its nuclear weapons a particularly difficult one. Had the political objective been restricted to punishing aggression and had United States military forces been capable of swift and strong reaction at the outset, there would have been no need to consider the use of nuclear weapons or the employment of a strategy which would have expanded the conflict beyond sensible geographic bounds.

The decision of the political and military leaders in Washington to exercise strategic restraint was a heroic one, because their field commanders, weak in tactical forces and knowing that powerful strategic weapons systems existed, were exerting strong pressure to employ a strategy of destruction rather than control. Bombing of the enemy's homeland might well have increased his will to fight and led to ever greater measures of destruction to bring him to terms. Military leaders, responsible for the strength, morale, and efficiency of their commands, generally want to employ every possible tool of war. If they are deprived of the weapons which permit control, they will demand and may have to be given the weapons which will spiral the world into a tornado of death and devastation.

It is, therefore, particularly important in the age of nuclear weapons that military strategy be subordinated to national strategy, that both military and nonmilitary means be available to persuade the enemy to stop fighting, and that military action should not be such as to contradict or override the effort to bring the enemy to the peace table. In theory, this has always been the case. But the awfulness of modern weapons has made it imperative that, henceforth, military strategy be really kept in its proper place.

It is important to note that the emphasis is not on keeping military men subordinate to civilians but on seeing that military strategy plays no more than a supporting role to national strategy. Paradoxically, it may be that in order to do this it will be necessary to elevate the military expert, for, in actuality, it is the civilian who has been responsible for putting so much emphasis on the military aspects of national security. It is the civilian who has been forcing the preoccupation with weapons by failing to provide alternate objectives and policies and by withholding the financial resources necessary to permit a flexible military strategy. There is good reason to believe that the military expert, if given a wide range of goals and ample means, would formulate alternate strategies. It is primarily because strategy has been rigidified by an extraordinary dependence upon large-scale nuclear weapons that it has come to dominate affairs.

If national strategy is defined as the art of developing and directing a nation's resources to the maintenance of national security, then military strategy must be the art of developing and directing military resources to the same end. Under this formula, the armed forces are to be used to achieve political aims rather than military ones. Indeed, no matter what the resources whether diplomatic or economic, psychological or military they should be directed toward the maintenance of national security and national aims, not simply military ones. Here is where we often make our great mistakes. By directing military force exclusively toward military ends, we let military strategy override national strategy. Thus, military action is allowed to contradict our economic and diplomatic efforts, and with today's weapons such a course can be a fatal one.

Some years ago, for example, the United States decided that one means of maintaining its national security was to rehabilitate Western Europe, building it up as an element of strength for the free world. To that end, it developed a national strategy which included employing its economic resources in giving economic aid to the Continent. At that time, American military strategy was based simply on attacking Russian cities from the air if the Soviet forces should move westward. Then, under fear of Soviet military aggression, the United States began giving military as well as economic assistance to the beleaguered governments of the area. It entered into diplomatic negotiations to form an Atlantic alliance in which it pledged itself actually to defend Western Europe. This was something new. The North Atlantic Treaty Organization was not designed for American forces to withdraw from Europe in the event of attack and then to return at leisure. It was designed to prevent Europe from being overrun and occupied. When the United States had a monopoly of atomic weapons, its military strategy was in line with its national aim of keeping Europe whole. The economic, the diplomatic, and the military elements of its national strategy were parallel.

On the day that the Soviet Union exploded its first atomic bomb, however, the situation changed drastically. Since that time, a military strategy based on using nuclear weapons has been out of harmony with our national aim of defending Europe. A nuclear war cannot preserve Europe; it can only insure its destruction and waste the economic resources we have poured into it. If, on the other hand, the United States does not use the nuclear strategy it has devised, it must, in the absence of other means, withdraw from Europe, and this, too, is contrary to its aims. If our military strategy can only produce destruction or withdrawal, it is dictating a national strategy of destruction or withdrawal, whether we plan it so or not. The only way the two strategies can be brought into line with policy is to provide weapons that will keep Russian bombers out of Europe and lend cover to NATO forces on the ground. Tactical air power, however, remains secondary in our military scheme of things, and short-range air defense missiles do not yet provide adequate means of keeping Soviet fighters and bombers out of Western European skies.

The United States has been driven by its choice of weapons to have one policy for peace and another for war. It has relied on the big bomb to preserve the peace in order that the fruits of its economic and military aid programs will not be lost, but the weapons it depends on to wage war have brought it uncomfortably close to a policy of destruction or withdrawal if the peace is shattered. Its military strategy is, in other words, designed primarily to achieve a military aim: a military victory over Russia through destroying her even at the cost of wrecking Europe.

The contradiction of policies is perhaps not quite so sharp as this dichotomy suggests. Nevertheless, the steadily increasing emphasis on firepower as a substitute for manpower in the NATO plan makes the situation perilously close to this description in essence, and it is a trend which should be reversed. If this country's NATO policies for peace and war are to complement each other, it is essential that our military, economic, and political strategies are made parallel so that they do not work at cross purposes. Only in this way can military strategy be a true servant of national policy. Our military plan in Western Europe should aim first at deterring war, and, failing this, at prosecuting that war in such a way that we can achieve a favorable settlement with a minimum of damage to a healthy Europe. We must not let our preoccupation with weapons of mass destruction make us frame a military strategy out of harmony with our ultimate objectives and out of keeping with the national strategy by which those objectives are to be pursued.

It may be argued that it is oversimplifying current American military strategy to imply that it is an all-or-nothing concept. The United States certainly possesses air, sea, and ground forces; it has some atomic weapons of relative precision as opposed to those of mass destruction, and it still retains a certain capability in conventional warfare. But before it can be maintained that we have the capacity to wage more than one type of war, several questions must be raised and answered. It is necessary to know how quickly the United States can put its forces for limited war into action and sustain them. It is essential to know if the military establishment is prepared logistically to shift from SAC and total war to infantry and brushfire conflicts. It must also be determined whether our strategic planning for unlimited war has precluded us from exercising the control necessary to keep the current conflict limited. The answers to these questions necessarily turn on logistic capabilities, for armed forces as instruments of control are of no use unless they can be quickly transported to the point of conflict and sustained long enough to permit them to stamp out brushfires and regain control.

One of the influences which modern weapons have exerted has been to subordinate the role of logistics to strategy to a dangerous degree. This may clearly be seen by examining the strategic and logistic situation in which the United States military establishment finds itself today. American strategy is designed primarily either to maintain the peace or to annihilate the enemy in war, and the main instrument to carry out both of these functions is the Strategic Air Command. Under the circumstances, SAC has correctly been given priority on our military air transport system, for, even if the big bombers are not called into action in small war situations, ample air freight must be held in reserve for it in order that it can retain its full capacity as a deterrent and retaliatory force. The question is whether under conditions which place such a heavy emphasis on SAC, our Military Air Transport Service, known as MATS, has been designed and can be employed to implement any other strategy especially one designed to control the enemy's actions by conventional means. The answer is that MATS provides a dangerously slim margin of safety because our strategic plans, being based so heavily on the all-or-nothing concept, do not provide for unforeseen emergencies. MATS is sufficient only to handle the situations which our strategy conceives as normal.

This strategy is not well geared to the politics of an age in which explosive nationalism and creeping Communism keep the international situation in constant turmoil. In the past few years, the Middle East alone has provided ample demonstration that peace is not to be kept by threats of massive retaliation and that wars are not to be fought and won with nuclear bombs. Again, the Anglo-French-Israeli attack on Egypt is a useful example. When the Suez crisis first arose in the summer of 1956, the French and British were logistically unprepared to move forces into Egypt to counter Gamal Abdel Nasser's seizure of the Canal. They had no choice but to let the situation deteriorate while they gathered strength in the Mediterranean for a show of force. Military action, if it was to achieve the objectives of restoring the Canal to international usage and of cutting the Egyptian dictator down to size, had to be swift and certain. The attacking powers were faced with the necessity of presenting the world with a fait accompli if they were to get away with the use of force against an Arab state. When it came, however, the attack was frustratingly slow, for, despite the weeks of preparation, the French and British were unable to land and support their forces by sea and air in numbers and speed sufficient to meet the situation. The Canal was effectively blocked by the Egyptians while the attacking forces were beating their way overland to reach it; world opinion was given a chance to mobilize in behalf of the Egyptians, and the Soviet Union was given time to fabricate its own disruptive plans.

The United Nations at this juncture called on Britain, France, and Israel to withdraw, and the United States decided that it was essential to get the invading armies out of Egypt as expeditiously as possible in order that Soviet volunteers would have no excuse to move in. The objective was to control the situation, to get the conflict back on a nonmilitary basis, and to keep out alien military forces which might bring on a war of greater scope and intensity. To that end, the creation of a United Nations Emergency Force was proposed and accepted. Time was of the essence to quench the smoldering fires, and, since air transport was the most rapid method of putting the Emergency Force on the scene, the United States offered its own aircraft on a limited scale to fly the Force to a staging area short of Egypt.

Given the basic premise that it was wise to arrest the conflict, this was a logical sequence of reasoning, and the plan worked reasonably well. The question at issue is, however, whether the United States could have done more along this line if the need had arisen for its doing so; whether, for example, it was prepared to fly the Emergency Force all the way to Egypt if serious fighting had continued there, or whether it could have offered more planes without unduly reducing the effectiveness of its own retaliatory forces. Certainly, if Soviet volunteers had begun to move in, the leisurely arrangement of flying the Emergency Force to points short of Egypt and then transhipping it to the scene of conflict would have been inadequate. Under such conditions, the choice before American policy-makers would have been whether to attack the problem at its source by dropping bombs on Moscow (and thus bring on the very war we are trying to avoid), or to fly troops of its own into the Middle East, either to limit the conflict geographically or to prevent war altogether by facing the Soviet volunteers with a superior force. The choice would have been determined by the extent of our logistical capacity, and as far as MATS is concerned, the evidence is that this was limited to the transport of no more than one division in size.

If the United States is therefore not to be forced into total war to achieve limited objectives, it must have forces capable of rapid transportation to the trouble spots around the world. The longer small wars are allowed to simmer, the more likelihood there is that they will boil over; and speed, therefore, is essential to the operation of any theory of limited war. The United States, separated by two great oceans from the major arenas of potential conflict in the world, has a special need for ample air and sea transport: the first to set the spearhead of its troubleshooters down at the scene immediately; the second to follow up with reinforcements and supplies.

The Lebanese crisis of 1958 demonstrated how narrow our margin of safety was in this respect. Marine battalions stationed with the Sixth Fleet in the Eastern Mediterranean were permitted to make an unopposed landing. The Army, having scraped together every bit of air transport on which it could lay its hands, then airlifted a few thousand troops from Western Europe. Had serious fighting developed in Lebanon, this meager reinforcement would have been too small and too late to restore peace and order with the efficiency and speed required by the exigencies of Arab nationalism and Soviet intransigence. Had the crisis occurred in the Persian Gulf or in Southeast Asia, where the Sixth and Seventh Fleets could have provided no overnight assault forces, immediate military action of any kind would have been impossible. Twice in the Middle East the United States has gotten away with its slim margin of air transport; in other areas and at other times it may not be so fortunate.

There is every likelihood that situations will develop in which we will not be able to exercise the control necessary to keep the conflicts within reasonable bounds, because weapons of mass destruction have influenced our strategy in a single direction and caused us to underestimate the logistic capabilities necessary for flexibility. The science of logistics is concerned, among other things, with movement with the transfer of men, equipment, and supplies from one point to another. What has been said of air transport applies in varying degrees to every phase of this problem, to every medium in which transportation operates. The United States should, therefore, be making every effort to increase the speed and carrying capacity of its military transport by land and sea as well as by air, and it should be examining the ways in which nuclear propulsion can be harnessed to this effort.

In assessing the influence of weapons on modern strategic thought, it becomes evident that mobility, as it pertains to weapons systems and logistics alike, has been subordinated to firepower, and that it must be restored to its historical place of parity in military doctrine. Since nuclear power has made the offensive so vastly superior to the defensive, a highly mobile weapons complex is essential to the retention of our retaliatory capabilities. Keeping pace with this requirement demands an integrated system of strategic thought in our military establishment i.e., less concern with the separate components of air power, sea power, and ground forces, and far greater emphasis on the logistical problems which a highly mobile weapons complex and an explosive international situation have already raised. Unfortunately, the new weapons of mass destruction have inflicted us with a kind of myopia. We have become so shortsighted as to plan our strategy in terms of an all or-nothing policy when in fact, more than ever before, we should be concentrating on the middle ranges of conflict and framing our strategies, both national and military, with a view to controlling the situation in order to prevent it from degenerating into all-out war. Control requires flexibility; it requires an assortment of aims, of means, and courses of action.

The strategist in the nuclear age has, therefore, a staggering task. He must first of all frame his plans so that they will be consistent with a wide range of national strategies, all of which are aimed at achieving the objective of survival and the maintenance of the nation's interests. For this task, he must provide himself in advance with a variety of tools men and machines, communication and logistic techniques, brain power and imagination. With such instruments at his command, he will have the opportunity to restore strategy to its proper level, to lift it from the depths of mass destruction to the higher plane of persuasion, pressure, and control. The strategist's challenge is to provide that intelligent direction which is the essence of strategy and which alone can save civilization.

9

Impact of Nuclear Testing

Introduction

A new round of speculation began as to how far India will adhere to the doctrine and exactly how it will take shape. Despite its articulation of "minimum deterrence, " there is a good deal of room left open in terms of specifics. In addition, there is still scope for revisions once the draft doctrine is introduced into parliament for discussion. Whatever the outcome may be, there is little doubt that India is attempting to confront its most critical dilemma at this juncture, that is, how to project a convincing yet non provocative and economical nuclear posture. This search for the right mix of policies goes to the heart of Indian thinking on the country's new status. As such, it offers a glimpse into not only emerging conceptions but also earlier perspectives and how they might be reconciled or reconstituted. Thus it is important to look at India's strategic practices and past positions for any important insights.

New Draft Nuclear Doctrine

Although some expressed impatience with the pace of Indian strategic planning, both in the United States and among sections of the strategic elite in India (albeit for different reasons), prior to the draft doctrine being made public, the doctrine as stated has hardly been welcomed by Western analysts. Among the factors that could have motivated the new Indian draft doctrine, the U. S. attempt to draw out India was one likely catalyst. Clearly, the United States was trying to steer India toward as muted and restrained a posture as possible. But India has tended to be much more responsive to domestic elite conceptions of security interests than to outside influence. Thus U. S. pressure was not likely to be heeded, in part because it has never been easy to square U. S. nuclear profligacy with appeals for nuclear abstinence and restraint in the developing world. In addition, no convincing evidence has ever surfaced to suggest that the "bean-counting, " arcane, and even incendiary strategic discourse and postures associated with the U. S.-Russian nuclear relationship are somehow superior to the virtues of a version of strategic ambiguity coupled with a concerted policy of reassurance, which India appeared to be adhering to subsequent to the tests. But nuclear doctrines themselves tend to take on a life and logic of their own, and it is not entirely surprising if the Indian draft looks very much like such doctrines elsewhere, not some unique Indian version.

India's current nuclear stance seems to be designed to leave little doubt that the country is serious about developing its nuclear weapons. The rather explicitly belabored draft doctrine would seem to be the near antithesis of the prior position. Indeed, the draft seems aimed at countering the earlier vagueness. Yet even though the document is unambiguous on strategic intentions, it is much less specific on the size and design of the so-called minimum deterrent. The time period for developing the deterrent appears to be long term up to thirty years, according to some of the doctrine's architects. Thus the elements of ambiguity have not entirely evaporated.

In giving practical shape to the contours of India's nuclear posture, the political decision makers who are placed Janus-like at the intersection of the international and domestic spheres will have to heed both internal and external forces and strike a balance. In a resource-poor country such as India, they will also have to make a careful tradeoff between military expenditures and economic priorities. In doing so, they are most likely to fall back on the centrist and accommodative tendencies of Indian politics and fashion a broad consensus at the elite level that tends toward gradualism and symbolism, buttressed by the declared nuclear doctrine and capability. Even if significant changes have occurred in the nuclear realm, prospective policy is still likely to have important predictable elements. In putting forth this perspective, I consider the main trends in strategic thinking over time, focusing particularly on the post-1994 period. It has seen an unprecedented flurry of international negotiations on nuclear issues, which in turn has deeply impacted India's own strategic practices, in many ways for the first time since Pokhran I in 1974. Taking the draft nuclear doctrine as the sum total of strategic thinking would seem to involve losing sight of the larger picture.

Strategic Schools of Thought

There is disagreement over whether this implies a critical lack of strategic thinking per se. The varying strands of opinion may be surmised as follows:

1. Strategic thinking is nearly absent in Indian tradition and is ad hoc at best.
2. It is clouded by normative considerations.
3. It is a function of strategic compulsions that India does not have.
4. It is discernible in some sectors and is based on strong historical tradition.
5. Strategic thinking, along with an emphasis on national security, is growing in India but needs to be eschewed.

To be sure, these statements are approximations, but it is possible to identify some of the adherents of the positions. The first position has been most forcefully put forward by George Tanham, who posits that Indian thinking is informed by multiple factors, particularly geography, history, culture, and British rule, none of them suffused with enduring strategic content. He finds that the strategic tradition has been woefully inadequate in Indian history, although his reference to Kautilya's mandala concept as the basis for current thinking in New Delhi might suggest the obverse. Kautilya's Arthasastra has long been considered one of the earliest treatises of classical realism, elaborating dictums all too familiar to later students of Niccolò Machiavelli. Tanham 's forays into the cultural sphere led him to make some deterministic findings on such subjects as the effects of the Hindu notions of time and planning, which are open to challenge. But in the extremely difficult terrain of strategic culture, what is required are more systematic interpretations before such observations can be simply admitted or dismissed. In a more recent essay, Tanham sees the potential for strategic theorizing but is not entirely convinced and tends to hedge what he sees as a situation "in flux."

Tanham's views are to some extent shared by key political decision makers and defense analysts such as Jaswant Singh and K. Subrahmanyam, but although the latter two rue the lack of institutional structures in place for coordinating national security policy, the former is relatively unconcerned. Subrahmanyam 's position may be closer to the second school of thought, which while decrying the poor state of strategic thinking in the country, blames this outcome on excessive moralism and normative concerns in Indian foreign and security policy rather than a historical or cultural inability to think in strategic terms. Here, the Nehruvian paradigm with its seeming neglect of realpolitik concerns is viewed as having led India to hold moralistically, normatively, and ideologically charged rhetorical positions in international forums that ultimately could not be supported and that others came to view as even hypocritical. This school of thought would have the most adherents among Indian policy analysts today.

An innovative third viewpoint based largely on structuralism notes that the development of strategic doctrines is intimately related to the functional need for such in a country's historical trajectory. The history of the war-prone European states thus produced the most elaborate set of security doctrines. This idea may be viewed as a variation on Charles Tilly's classic proposition, which may be paraphrased as "war makes the state, and state makes the war." Although this rendering may be intellectually and historically compelling, the accompanying view that India does not face major strategic compulsions in the contemporary period, particularly from China, is not likely to be shared by many analysts.

The fourth perspective, which argues that India indeed has a strong tradition of strategic thought that is missed by outsiders such as Tanham because they are searching for the wrong things in the wrong places based on Western models (like open, written documents and coordinated institutional structures), has some currency. However, this thesis remains rather undeveloped, although in part it resembles the critique that has emerged from Southeast

Asia regarding strategic practices and norms in the East versus the West.

The view that strategic analysis is increasing in India but should be eschewed is offered as a critique of India's current nuclearization. Much of this critique appears to be based on the rise of the so-called national security state in the United States during the Cold War and the abuse of national security in the service of secrecy, intervention, and spiraling defense budgets. Adherents worry that if Indian politics becomes dominated by security concerns, "piercing the veil of national security" would become increasingly difficult, which would have an unhealthy impact on democracy. This view is not yet widespread.

In light of the above discussion, a rather fundamental point may be worth stating. In assessing the state of India's strategic thinking, two factors need to be kept in mind. Such thinking involves definitional questions, as well as relative or comparative aspects. In politics, strategy is critical, and survival depends on it. Strategic thinking can take different forms, but the simplest way of conceiving it is that a country pursues its self-interest (however it may be defined). In international politics, this means that a country acts in ways that protects its security. Strategic action can be informed by strategic vision or military strategy. If the latter is taken as the leitmotif of strategic thought, then India in comparative terms could be found lacking in a tradition. But that would not hold true if the former is viewed as a hallmark of strategy.

It may be argued, for example, that in the post independence period, Prime Minister Jawaharlal Nehru's approach was well crafted to protect India's autonomy in a postcolonial, bipolar world by emphasizing a rule based system, which could work to the advantage of independent, smaller powers, rather than a force-based international one. Nehru's rejection of the specific, militarized ideological Cold War and the two-camp realpolitik approach has in many ways been misinterpreted as a simultaneous rejection of realist thinking narrowly and strategic thinking more broadly. Under the circumstances, Nehru's nonalignment and multilateralism could be understood more in terms of strategic vision than strategic naivete. In Nehruvian terms, India's nuclear program fit into this broader scheme for national autonomy.

Nuclear Program

There has been no dearth of research and writing on India's foreign and security policy since 1947 or, for that matter, since 1974, but sustained discussion and thinking on the Indian nuclear issue was relatively dormant and occurred only sporadically until 1995. This situation reflected not only the lack of public information available on the nuclear program but also the generally widespread perception that nuclear weapons did not figure centrally in India's approach to the world.

The nuclear program itself proceeded apace from government to government, with hardly a break since the establishment of the Atomic Energy Commission (AEC) in August 1948. The key architect of India's atomic program, Homi Bhabha, enjoyed a close relationship with Nehru, who gave the former wide latitude to pursue this effort. Bhabha had recommended to Nehru that the administration of this program be entrusted to a small high-powered body such as the AEC, answerable directly to the prime minister without any intervening link. This setup laid the groundwork for an atomic energy program that enjoyed top support and remained insulated to a great extent from outside critical review.

Nehru remained opposed to nuclear weapons publicly and used this issue to attack U. S. and Soviet policies. Whatever his true persuasions were regarding a nuclear weapons program for India, there is no mistaking his enormous commitment to the development and advancement of Indian science and technology, particularly those technologies seen as the critical bridges to self-reliance. In the postcolonial environment, to a great extent, advanced scientific capability became identified as a measure of national pride and autonomy at both the elite and popular levels, which also helped to sustain fairly uncritical support for the nuclear program.

However, after the 1962 Sino-Indian war and the public declaration by the Chinese in 1964 of their intention to develop atomic weapons, followed by a fission test that same year and a second test in 1965, India became embroiled in a wide-ranging debate about appropriate Indian nuclear strategy, but without the benefit of Nehru's views (he had died before the Chinese went nuclear). Then AEC chairman Vikram Sarabhai and the director of the newly set up Institute for Defence Studies and Analyses (IDSA), Major General

Som Dutt, were among those unconvinced that India required nuclear weapons. However, in a speech to the International Atomic Energy Agency after the second Chinese nuclear test, Bhabha stated that it would be difficult to follow a policy of restraint given the introduction of nuclear weapons in the neighborhood. Soon after the Chinese thermonuclear test in 1967, a scientific effort was begun to study the nuclear design of an explosive.

Nuclear Threshold Stalemate

The political debate over India's nuclear stand seemed to come to a rest by 1970, with the decision not to sign the Nuclear Nonproliferation Treaty (NPT) (open for signature from 1968 to 1970) but not to join the nuclear club either. The official public position, keeping the "nuclear option" open without committing one way or the other, provided Indian decision makers political and strategic space for the sometimes contradictory policies that followed. The notion of strategic ambiguity allowed India to pursue multiple objectives, including global disarmament, dual-use technological capabilities, a fairly autonomous foreign and security policy, a substantial defense, and an elevated status internationally.

This threshold period is best characterized as one of "technology demonstrations, " which may be interpreted as a form of strategic posturing with strong symbolic significance for the credibility of the latent nuclear option but was also marked by public self-restraint. The establishment of the integrated guided missile development program (IGMDP) in 1983 and the successful test firing of the short-range ballistic missile Prithvi in 1989, as well as the intermediate-range Agni in 1994, were viewed as critical steps in maintaining the option vis-à-vis Pakistan and China, respectively. India's technical achievements on the civilian side in the space program, such as the first operational launch of the polar satellite launch vehicle (PSLV) in 1997 and the successful indigenous Param computer project using parallel processing architecture to match the speed of modern supercomputers after the U. S. denied the sale of the Cray supercomputer, underscored India's dual-use capabilities.

Over time, India's threshold position, symbolically shared by Pakistan and Israel, came to be perceived as a nearly semipermanent state of being. Such an ambiguous status gained in import by being theoretically formalized as non weaponized deterrence or recessed deterrence. Moreover, India basically adhered to the principles of nonproliferation that underlie the NPT while not formally committing itself to do so. But technically, it continued to fall outside the confines of the NPT, and in the eyes of the United States in particular, India's image was that of a revisionist state destined to be at odds with the United States, which saw the NPT as representing the core of global nonproliferation efforts.

As a threshold nuclear state, India neither tested nor deployed nuclear weapons; nor did it transfer sensitive nuclear technology or train nuclear experts from other countries. Although there was a broad consensus for protecting the country's nuclear option among the scientific, military, and political elites, there was no significant lobby for "going nuclear." During this time, neighboring China was steadily advancing the size and sophistication of its nuclear arsenal as well as providing critical input into Pakistan's nuclear and missile programs. Meanwhile, the United States and the Union of Soviet Socialist Republics (USSR) had continued their intense competitive buildup as well. Given that military planners assess capabilities rather than intentions when making strategic choices, India's past restraint could be viewed as exceptional. Yet, India's self-imposed restraint received little acknowledgment, either from the United States or the international community. This would prove to be an important factor as India moved to cross the threshold in May 1998.

Turning Point for the Comprehensive Test Ban Treaty

Maintaining the threshold status appeared to have widespread support inside and outside government, even as India entered the difficult decade of the 1990s. Indeed, in 1991 India faced twin crises of historic proportions in both the security and economic realms. India's carefully crafted relationship with the USSR (of nearly two decades) lay in disarray in the stunning wake of the Soviet collapse, leaving Indian foreign policy adrift. A financial crisis reduced India to barely a fortnight's worth of foreign exchange reserves to cover its imports amid growing skepticism regarding the limits of the Nehruvian state-led economic development model. Thus two paradigms that had reigned supreme in the post independence period were coming undone at exactly the same time. The task of negotiating these crises fell to the coalition government led by the Congress Party (which was just short of a majority) under P. V. Narasimha Rao, elected in 1991 A combination of economic liberalization, improvement of ties with the United States, a "look east" policy toward Southeast Asia, all suggesting a new and more variegated Indian external approach, served to put India on sounder footing than might have been expected under the circumstances.

Yet India's improving relations with the United States, fueled in part by greater convergence of economic interests, could not contain the strategic divergences that persisted. The Clinton administration's pursuit of nonproliferation in one of the most intense efforts since the NPT was first conceived made it inevitable that this issue would not disappear or even recede. In this context, the NPT Review Conference in May 1995 and the Comprehensive Test Ban Treaty (CTBT) negotiations in 1996 became critical junctures for India to reconsider the "threshold" policy publicly. India's decision to veto the CTBT in Geneva in 1996 placed it in the unenviable position of having to reject something that it had been on record as championing since 1955. Indeed, Nehru had successfully instructed the Ministry of Foreign Affairs to have India be the first to sign the Partial Test Ban Treaty in 1963. India had cosponsored the CTBT resolution twice with the United States in the UN General Assembly in 1993 and 1994, without engendering opposition from the Indian scientific, military, or policy communities. Thus India's about-face on the CTBT needs explanation, since this policy shift represented in important ways the beginning of a break with traditional strategic thinking and practice.

The outcome of the NPT Review Conference in May 1995, extending the treaty indefinitely and unconditionally, came as a shock to most Indian analysts. The expectation was that there was enough support from key middle-power developing countries such as Mexico, Venezuela, Egypt, and Indonesia to extend the treaty for only another twenty-five years and even make that conditional on firmer commitments by the nuclear weapons states to move toward eliminating their arsenals according to Article 6 of the NPT. This view was clearly a miscalculation on the part of India, given the vulnerable positions of countries such as Mexico, which was trying to weather one of its worst financial crises, and Egypt, which was highly susceptible to U. S. pressure. Indications that it would be difficult to sustain a push for disarmament came clearly only toward the end, however, for example, with the inability to get support for secret balloting. For Indian decision makers, the NPT Review Conference was thus the dawning that India's traditional agenda disarmament coupled with nonproliferation was unlikely to get serious results. India learned an object lesson from the NPT extension: the ability of the non nuclear states to influence the nuclear weapons states by "holding their feet to the fire" was fast eroding. Indian diplomacy began to be viewed as increasingly out of step with changing international realities, and at the same time, discussions within India became increasingly self-critical. This critique found its fullest expression during the CTBT negotiations. Indeed, until the NPT extension, the CTBT had not come under the scrutiny of public debate, but after May 1995, the treaty began to be looked upon in a different light.

A number of concerns converged to impinge on India's CTBT decision, not all directly related to the intricacies of the negotiations at hand at the 1996 Conference on Disarmament in Geneva, including some that had been simmering for a long time. During the intense domestic debate occurring in 1996, Indian diplomacy on security was faulted along a number of lines: the inability to change a persistent outside tendency to equate India and Pakistan and to not take Indian concerns vis-à-vis China more seriously; the failure to gain external acknowledgement of India's broad restraint in the nuclear arena; the inability to put forth national interest in strategic, realpolitik terms as a result of being boxed in with past "moralist" baggage; the possibility that the nuclear option for India would be closed "in effect" after the CTBT, since twenty-two years of non testing made it less "credible"; and the perception that the treaty was aimed specifically at India, since the other two threshold states had nuclear patrons, and India did not. All this was being assessed against the rising conviction that the nuclear weapons states would continue to retain their advantage indefinitely. Thus in January 1996, without warning India introduced "time-bound disarmament" as a condition for its signature. In light of these concerns, the introduction of such an increasingly unviable condition would suggest that it also served as an "exit strategy" for India. The entry into force clause of the

CTBT in the final stage of negotiations only reinforced the Indian perception that the treaty was indeed aimed at India and produced a profound domestic reaction that sealed the decision to veto.

The evolution of India's CTBT position also sheds some light on how domestic and international factors may interact in this arena. The period 1995–1996 witnessed three ruling coalitions in a row. In May 1996, the Rao coalition government lost power, followed by a thirteen-day interim government during the same month led by the Bharatiya Janata Party (BJP), which failed to secure enough support to actually form a government by consent. Subsequently, the United Front, a fourteen-party coalition with outside support from Congress, formed a government that managed to last from May 1996 to March 1997. Despite the wide spectrum of political leanings encompassed within these three ruling groups, the entry into force clause saw a closing of the ranks supporting a veto, which was seen as necessary to avoid being painted into a corner. In anticipation of this outcome, India had been willing to withdraw from the Committee on Disarmament and let it run its course without its participation, which turned out to be unacceptable, in particular, to China.

This unexpected turn of events at the Committee on Disarmament contributed significantly to the hardening of India's stand against the CTBT and strengthened those who favored a stronger posture, with India's chief negotiator Arundhati Ghose declaring that India would not sign the treaty, "not now, not later." The Indian veto was an important symbolic turning point in strategic posture, and with Ambassador Ghose 's speech citing national security concerns in a way not specified before, it provided the opening for a reduction of traditional normative discourse that had until then been central to Indian strategic thinking. In this context, the perceived international pressure via the entry into force clause left little credible space for those in favor of India signing the CTBT.

One point of near unanimity among the political parties in India during the CTBT debate was the need to continue to support India's nuclear "option." A key question related to just how to safeguard it, however, including whether testing was necessary or not in order to lend credibility to an option that had been kept open for twenty-four years without much public display. The veto at the CTBT was presented as evidence of India's "resolve" to maintain this position and not give in to international pressure. Indeed, in some quarters there was almost relief that by being "forced" to take such a strong action at the Committee on Disarmament, India was able to establish the credibility of the nuclear option without having to test. Since India has gone one step further and tested, the question of credibility has again come to the fore in a different fashion.

Strategic Thinking After the Tests

The question that may be posed at this stage is the extent to which the nuclear tests have fundamentally changed the strategic thinking in India and what it means for Indian security policy. Do India's dramatic tests mean a complete break with India's past strategic thinking? Prior to the tests, India had straddled a unique position on the threshold, with a blend of normative rhetoric on disarmament and some level of realpolitik action in its ongoing nuclear program. The draft doctrine's two focal points minimum deterrence and the credibility of the deterrent pull in opposite directions, and one challenge will be to resolve the two tendencies.

The Indian government initially announced elements of its post-testing posture, which included a no first use doctrine, minimum deterrence weaponization, a moratorium on testing with broad hints of signing the CTBT, and a willingness to participate in the on-again, off-again fissile material cutoff treaty (FMCT) negotiations in Geneva. The Vajpayee government maintained an ambiguous position on what it meant by minimum deterrence but provided some outline by way of declarations and overtures. This approach was consistent with India's past practice of taking a strategic position perceived as economical, nonprovocative, credible, and flexible; not spelling it out in detail; but assuming it enjoyed a fairly widespread consensus among the ruling and influential elites. This finely balanced act came under increasing pressure from a variety of quarters, and the draft nuclear doctrine suggests that this balance has been tipped, at least on paper.

In the past, Indian strategic elite or political decision makers have shown little support for following a fast track nuclear weaponization program. In the avid debate regarding the level of weaponization, the two levels most often discussed were either in the low range of 60–130 weapons, depending on the source, or in the high range of 300–400 warheads, which approximates the Chinese, British, and French so-called minimum deterrents. The triadic "survivable" nuclear force mentioned in the draft doctrine subsequent to the strategic review on long-term policy options would suggest that the lowest numbers mentioned are likely to be surpassed. The release of the draft doctrine has not fully resolved this question, though the issue now appears to have shifted to whether a low three-digit number would be better than a higher three-digit number. If the draft is taken as a statement of "general principles," as some members of the National Security Advisory Board (NSAB) have said, nothing definitive about the configuration of the nuclear force can be concluded.

Independent analysts such as C. Raja Mohan, K. Subrahmanyam, and Jasjit Singh, the director of the government-supported Institute for Defence Studies and Analyses, all of whom enjoy influence across political parties, seem to be forging a clear consensus against the higher numbers. The National Security Advisory Board, which was set up in November 1998 with twenty-seven members and Subrahmanyam as the convener, continues to be broadly representative. The NSAB complements the three-tiered National Security Council headed by the prime minister, which was established earlier.

The long-unanswered nuclear "demonstration versus deployment" question has been resolved toward the latter with the tests, and although the space for nuclear restraint has narrowed, there is evidence that India is still seriously searching for ways to reduce the danger from an overt nuclear position. One such proposal by some Indian analysts is de-mating, that is, keeping warheads separated from their delivery vehicles. Accidental or precipitous use of nuclear weapons would be inhibited by this posture, which stops short of full readiness. One of the world's foremost scholars on nuclear safety concludes that if this route is taken in India, it could become a model for other nuclear states. Another expert notes that India, and even Pakistan in the past, may have followed a "deliberate policy to maintain a de-alerted status" and a delayed launch procedure strategy to minimize accidents, which if continued in some fashion would address some of the most critical command and control problems.

There were indications that India did not plan to go down the path toward full-scale deterrence as practiced in the West, with the attendant intelligence, delivery systems, and communication support, the cost of which might prove to be prohibitive and derail its economic experiment in the process. Economic priorities have been at the forefront since 1991, and political parties of all types agree on the need for staying this course. On a number of occasions, Vajpayee stated that a secure command and control system was already in place for the Indian nuclear arsenal, thereby deflecting further preparation that could spiral out of economic control.

Others suggested that under conditions of minimum deterrence, a fully delineated nuclear doctrine was not needed. As a well-placed Indian expert pointed out, "We don't fall into the standard pattern of declared doctrines, specific weapons, delivery capabilities or force postures, " and he added that minimum deterrence would make the likelihood of a response to "aggressive acts" more credible, even without the doctrinal apparatus. Another well-known analyst from the army pointed to differences between the grand deterrence strategy of the United States and the minimum deterrence India is practicing, emphasizing that although the U. S. philosophy is complex and designed to achieve a large number of variable objectives, the Indian deterrent is required "purely as a defensive instrument to ensure that no outside power is tempted to coerce the country or initiate a nuclear strike in a conflict situation. Therefore the size numbers and capability range of the strategic forces need to be limited to deterring regional nuclear powers. "

India's minimum deterrence thinking is not dissimilar in theory to the principles followed by Britain, France, and China. In a comprehensive study of these three countries'; nuclear doctrines, Bruce Larkin notes the striking difference between their "answers" (in terms of numbers of missiles) to the following question and those of the United States and Russia: "In a world in which my forces could be attacked first, what is the smallest force that guarantees deterrence?" Larkin emphasizes that the logic of their choice is driven by the notion that "deterrence is not achieved by vast numbers, but by the credible capability to strike at all." China is believed to possess roughly twenty missiles that can reach the United States and approximately 300 nuclear weapons that can reach Japan, Russia, and India. U. S. commentators have referred to China's nuclear weaponry as a "bare-bones arsenal compared with the thousands of warheads still maintained by the United States and Russia. " It has been pointed out by other Western experts that China's minimum deterrent was designed to retain the ability, after a major attack by a nuclear adversary, to launch at least one or two missiles that could destroy a major city.

The learning curve based on the U. S.-Russian overkill of nuclear weapons acquisition seems to be steep when one considers the choices of China, France, and Britain. The experience of the second-tier nuclear powers could in turn have a similar impact on India. In a statement to the Indian parliament, Vajpayee made the following oblique reference to China, noting: "We are not going to enter into an arms race with any country. Ours will be a minimum credible deterrent, which will safeguard India's security. " Thus the draft doctrine does not seem to be country- or threat-specific.

In the post-testing environment, continuing talks between India and Pakistan sought to reduce the volatility of the new nuclear situation, culminating with the historic bus ride to Lahore in February 1999. The talks extended important agreements already concluded, such as the agreement not to attack each other's nuclear facilities. Immediately after the tests, both then Pakistani prime minister Nawaz Sharif and Vajpayee spoke of negotiating from "strength, " which suggested that they saw some sort of a deterrence paradigm operating. However, they also expressed serious concern that competition did not get out of hand, or "spiral. "

Subsequent to the Kargil conflict of 1999 where Pakistan sent armed insurgents into the Kargil and Drass sectors of Indian Kashmir however, the presumption that the worst fears of the spiral theorists would be more than moderated by a rational commitment to make deterrence work needs to be reconsidered in light of Pakistan's apparent willingness to engage in such risky behavior. It begs the question of just how the nuclear equation is being interpreted by Islamabad. That the Kargil intervention was launched just as the nuclear fear was being dampened through diplomacy such as the Lahore summit is particularly sobering. In this connection, the effect of the U. S. and even the Chinese reaction to Pakistan regarding Kargil has no doubt been to deflate Pakistani expectations about what it might accomplish behind the nuclear curtain. India's challenge will be to demonstrate restrained behavior vis-à-vis Pakistan.

Summary

For India, a period of flux in its strategic thinking seems to be in the offing. How far the nuclear tests and weaponization breaks with India's past tendency to fashion security policies that were convincing but economical is still an open question. Indian decision makers functioned within an ambiguous nuclear realm for nearly a quarter of a century. India's past strategic doctrine of ambiguity has in some ways led to the underdevelopment of strategic theory in Western terms. However, historically speaking, doctrine has often followed action rather than the other way around. For example, the doctrine of mutual assured destruction (MAD) was spawned by the recognition that the Soviet Union and United States could inflict unacceptable damage on each other.

It remains unclear whether the elements of lower risk and lower costs (domestic economic and international political costs) will predominate. The possibility that technological momentum gained with the nuclear tests and the "logic" of nuclear deterrence may push India into a more "maxi-mimalist" minimum deterrent position cannot be ruled out. To posit this as a likelihood based on the U. S.-Russian experience or organizational theory, however, is misplaced. These arguments would have to also then explain how political decision makers in India resisted the technological drive for testing for more than two decades. Moreover, in the current nuclear debate, the technical elite is not monolithically in favor of a rapid or large-scale buildup of capabilities. Indeed, A. P. J. Abdul Kalam, one of India's chief architects of its missile and nuclear program, has publicly stated that India could sign the CTBT without compromising the country's security. According to Kalam, "From the scientific and technical angle, it is our considered view that no further nuclear tests are necessary."

In the short to medium term, the exigencies of the military-economic tradeoff cannot be avoided by India. India's ability to absorb the costs of a triadic nuclear force remains a matter of contention, in part based on differing assumptions of the size of the nuclear force, projected economic growth rates, and the time horizon. Two examples are useful here. In the case of the United States, as early as 1964, Defense Secretary Robert McNamara had concluded that the existing arsenal of approximately 400 nuclear weapons amounted to a credible deterrent. The mindless growth in numbers of U. S. weapons to more than 70,000 and the staggering costs of more than $5 trillion stand as the worst possible case. Likewise, the Soviet Union collapsed under its military weight. But if the actions of second-tier countries are taken into account, particularly those of Britain and France, neither country would appear to have careened out of control because of weapons acquisitions.

Of course, security is a highly elastic concept that is open to wide interpretation. If Indian past behavior is any guide, the dominant tendency will be to craft an economically viable posture. India's party politics have historically pushed policies toward the center, and with the onset of serious coalition politics, this tendency will get even stronger. The coalitions are increasingly being made up of smaller, regional parties with policies driven by local interests. They will be forced to carefully consider the economic-military tradeoffs and their impact on votes.

Moreover, should India move toward signing the CTBT and participate in a successful prospective FMCT, these treaties together would impose qualitative as well as quantitative limitations on India's nuclear weapons development. In light of the nuclear weapons states already having conducted more than 2,000 tests and with the United States and Russia awash in plutonium and highly enriched uranium, an agreement by India to cap itself at a much lower level of capability and thereby for all practical purposes be locked into a position of genuine "minimum deterrence" needs to be seen as a historic step. It would then guarantee that the blind buildup by the current purveyors of nonproliferation would not be emulated, indeed, could not be emulated. Of course, these treaties would have the likely effect of freezing the status quo, which would in relative terms give India an advantage over Pakistan, China an advantage over India, and the United States and Russia an advantage over China.

In the past, India kept its so-called nuclear option open for nearly twenty-five years without exercising it. In a sense, the draft nuclear doctrine appears to have been charted in such a way that the shape of the nuclear capability is left open without closing off any options in the future. Thus, although the draft doctrine could be seen as rhetorically fairly ambitious, it still leaves the possibility in practice of a nuclear posture that reflects India's traditional aversion to risky and expensive military solutions.

10

Pakistan's Nuclear Parity with India

Introduction

Pakistan's twofold strategy was to develop a credible nuclear deterrent against India and to fight international pressure against its program. Smarting under years of punitive sanctions, particularly those imposed by the United States, in 1995 Pakistan assured the world that it was pursuing a policy of non weaponized nuclear deterrence. But the Indian nuclear tests of May 11 and 13, 1998, forced Pakistan to exercise its option of a minimum nuclear deterrent. Pakistan said that it was left with no choice but to respond to the Indian tests so as to restore the "regional strategic balance." One can also objectively establish that the initiation and maturation of the Pakistani nuclear program has hinged on India and Indian nuclear policy. In the 1998 post detonation phase, this pattern of correlation has continued and become intertwined with the determination of a minimum deterrent for both India and Pakistan.

In the 1950s Pakistan became apprehensive of Indian nuclear acquisitions. The first Indian prime minister, Jawaharlal Nehru, stated of India's nuclear policy: "Indian scientists will use atomic force for constructive purposes. However, if India is threatened, she will inevitably try to defend herself by all means at her disposal." Such statements induced Pakistan to monitor (and develop fears about) the Indian nuclear program. After the 1962 Sino-Indian war, when India started developing its nuclear weapons program in earnest, Pakistan initiated a process of espionage and clandestine acquisition of nuclear technology and design. With the 1971 Indo Pakistani war and subsequent nuclear explosion by India in 1974, Pakistan pursued nuclear weapons in a more serious and organized fashion. Since the mid-1980s, two parallel processes can be discerned. First, a series of declaratory statements emanated from the establishment confirming Pakistan's "capability. " Second, a spate of sanctions imposed on Pakistan continued in one form or the other until May 1998 and afterward.

The May 1998 Indian tests brought Pakistan to the point at which it finally "jumped into the flames with courage. " Shortly afterward, Pakistan briefly "delinked" itself from Indian nuclear policy, particularly on the question of the signing and ratification of the Comprehensive Test Ban Treaty (CTBT), but this delinkage was only an ephemeral phenomenon. The Indian Agni 2 missile test on April 11, 1999, prompted Pakistan to test its own missiles the Ghauri 2 and Shaheen on April 14 and 15, 1999 Knowing that India was going to test, Pakistan delayed its own tests so India would bear the brunt of international opprobrium. Pakistan's statements afterward, that the retaliatory missile tests by Pakistan were merely a national security requirement, did not sound totally plausible. Similarly, although Pakistani leaders had repeatedly announced that they would not link their decision to sign the CTBT to a similar decision by India, they became nervous following the collapse of the Vajpayee government and started dropping hints about relinkage.

Thus Pakistan finds itself in a highly fluid situation, whereby its own nuclear policy keeps evolving on an ad hoc basis vis-à-vis India. Delinking and relinking nuclear policies, doing belated homework for the elaboration of doctrines, and keeping its defense decision making in flux have catapulted Pakistan into uncharted territory where it cannot make completely independent decisions.

After acquiring an overt nuclear status, Pakistan faces myriad challenges that can be broadly categorized into four areas. First, Pakistan has to calibrate its policies and responses to highly fluid and uncertain Indian policymaking dynamics in the nuclear and missile fields, complicated by the fragility of Indian coalition politics.

Second, Pakistan has to remain engaged with the United States on nonproliferation benchmarks to avert punitive sanctions. Third, it has to grapple with the imperative of conducting internal consultations to prepare a blueprint of a nuclear doctrine, an elaboration of command and control systems, and a definition of minimum deterrence. Fourth, Pakistan must mount a gigantic effort to keep its economy afloat and strengthen it to sustain its minimum nuclear deterrent. Discussions of these four problematic areas overlap in the sections to follow.

Jumping into the Flames with Courage

India's nuclear tests resonated with a deafening force in Pakistan, which was consumed by outrage at India and at the Pakistani government's ambitions to prove its own nuclear capability. The leader of the fundamentalist Jamaat-i-Islami (Islamic Party) called for an "immediate" response. Abdul Qadeer Khan, the head of Pakistan's nuclear program, said "it is a question of hours and not days" for Pakistan to carry out nuclear tests. The images of Pakistani children burning Indian flags and youngsters shouting "Qadeer Khan bumb nikalo!" ("Qadeer Khan take out the bomb!") reflected what the masses wanted then Prime Minister Nawaz Sharif to do. They were more concerned with encouraging a "tit-for-tat" response to India than with the dire economics of sanctions. For them, sanctions were nothing new; the United States had repeatedly imposed them to punish Pakistan for its nuclear program. Extremist elements in Pakistan had pressed for a test since the late 1980s, soon after Pakistan was known to possess bomb-making capabilities. The majority favored maintaining the status quo, that is, an ambiguous nuclear posture. However, an overwhelming consensus supported going ahead with a test if India tested its bombs.

Though Prime Minister Sharif held the biggest public mandate in Pakistan's history and faced no direct threat to his government in May 1998, he was acutely conscious of the fact that sitting tight on the nuclear test decision would erode his popularity drastically. Former prime minister and opposition leader Benazir Bhutto's prediction that "either Pakistan detonates a nuclear test device or it risks a possible war with India over Kashmir" was endorsed the very next day by Indian home minister Lal Krishna Advani's threatening statements on Kashmir. With all eyes focused on Sharif and his expected announcement that Pakistan would carry out nuclear explosions, it was difficult for him to bargain over international aid and incentives that the Pakistani government considered "too little, too late."

In a joint telephone call before Pakistan's tests, British prime minister Tony Blair and U. S. president Bill Clinton unsuccessfully sought to persuade Sharif not to go through with threatened nuclear tests. A U. S. delegation headed by Deputy Secretary of State Strobe Talbott flew to Pakistan on May 13, 1998, but came away without any evidence to support Clinton's hopes that Pakistan was moving away from conducting a test. The delegation did make it "very, very clear" to Pakistan that it would face horrible consequences if it chose to conduct its nuclear test. This visit was deja vu for Pakistan. Secretary of State Henry Kissinger had flown to Pakistan in 1976 to deliver a similar warning that he would make "a horrible example" out of Prime Minister Zulfikar Ali Bhutto and Pakistan if Islamabad went ahead with a nuclear weapons program.

U. S. officials had not offered any concrete proposal but rather suggested "possibilities," such as waiver of the 1985 Pressler amendment (which conditions U. S. aid to Pakistan on a presidential certification that Pakistan does not possess nuclear weapons); delivering F-16 fighter aircraft that Pakistan had purchased from the Bush administration but that had been held up under the Pressler amendment; and possibly some debt retirement. The Clinton administration appeared to focus primarily on preventing a Pakistani nuclear test rather than on identifying tangible measures that could alleviate Pakistan's broader security concerns. From Islamabad, it appeared that the United States was belatedly offering a warm hug, which Pakistan could not reciprocate.

Prime Minister Sharif also charged that the United States had been discriminating against Pakistan on the basis of religion, arguing that the U. S. government had "ignored" India's nuclear activities and sanctioned Pakistan's because "we are an Islamic country." The Organization of the Islamic Conference (OIC) was quick to respond by expressing its solidarity with Pakistan after the Indian explosions. If the United States had pushed Pakistan against the wall by intensifying existing sanctions, it would have risked

Pakistan's drift toward its "Islamic brethren," raising the specter of nuclear technology transfers to "rogue states."

Despite China's clandestine nuclear assistance to Pakistan, the former's ongoing détente with India in the last decade had been a source of anxiety for Islamabad. Thus, even as Indian defense minister George Fernandes cited China as India's prime enemy, China condemned the Indian tests, and the Indians charged China with "hypocrisy," Pakistan's fears were not assuaged as it grappled with the new "explosive" threat from India. Pakistani foreign secretary Shamshad Ahmed's visit to China immediately after the Indian tests was probably designed more to assess China's potential reaction to an escalating South Asian crisis than simply to push its diplomatic offensive against India. Pakistani calculations were that as Russian assistance to India increased, China would be more pivotal for Pakistan than the jubilant Arab states. Pakistan wanted to know where China stood as Islamabad charted its response to India and prepared contingency plans for any ensuing crisis.

Indian home minister Lal Krishna Advani affirmed India's "resolve to deal firmly and strongly with Pakistan's hostile designs and activities in Kashmir" and solicited a more "pro-active" posture against Pakistan.

Coming on the heels of Prime Minister Vajpayee's warnings that India could use nuclear weapons in cases of "aggression from outside," this statement was viewed in Pakistan as a threat. Indian analysts initially welcomed the prospect that Pakistan would conduct a nuclear test of its own in response to India's explosions. However, the Indian government was apparently rattled by the suggestion in the New York Times that by refraining from testing, Pakistan could get its debts written off. India's renewed vows to proceed with its nuclear weapons program despite sanctions and to develop a new longer-range missile capable of delivering nuclear warheads may well have been intended to push Pakistan into carrying out a nuclear test. Even more dangerous were Advani's threatening statements, which were read as promising a nuclear strike on Pakistan if Islamabad continued to pursue its "anti-India" policies. Leaders in Islamabad believed the world community had not fully grasped the implications of the Indian tests and of the provocative statements that had been issued from Delhi in their aftermath. Even more incomprehensible to Pakistanis were assertions from the U. S. State Department advising Pakistan that "they would be far, far better off if they chose the diplomatic road, the high road. " This was not the road chosen.

Economic Situation

In the midst of a serious balance of payments crisis, in 1998 Pakistan's fragile economy was heavily dependent on International Monetary Fund (IMF) support and was on the verge of defaulting on its international debt obligations, despite the fact that the Overseas Private Investment Corporation (OPIC), the Export-Import Bank, and the Trade Development Authority (TDA) had restored their operations in Pakistan and the IMF had expressed cautious optimism about the success of the package launched a year earlier. A nuclear test by Pakistan that accelerated an arms race with India could easily lead to additional military spending commitments far beyond what its economy could sustain. Statements by then finance minister Sartaj Aziz suggested that Pakistan was aware of this risk. In contrast, Indian analysts claimed that U. S. sanctions (excluding the lending by international financial institutions) would amount to a mere 1 percent loss of gross national product (GNP). Some U. S. legislators and opinion makers claimed that the sanctions would hurt the United States more than India. The European Union appeared resolved not to isolate India; France and Russia resisted sanctions against India; and the Europeans seemed ready to grab any Indian business from U. S. companies frozen out by U. S. penalties.

The Pakistani government seemed to understand that sanctions that might be only a pinch for India could bite deeply into its own economy. Even though the decision to test went beyond economics, the impact of sanctions was worse than Pakistani decision makers had expected. The suspension of the IMF package, termination of lending by the World Bank and the Asian Development Bank, and the mounting debts created arguably the worst financial crisis in Pakistan's history. Euphoria over the bomb was replaced by a sobering realization. The freezing of foreign currency accounts by the government hit the Pakistani expatriates hard and eroded confidence in Pakistan's overall financial viability. Nervous Pakistani economic managers fended off growing debate about an impending implosion and a failed state.

Then the United States came to Pakistan's rescue. In early 1999, the IMF approved a package of about $1.69 billion to support structural adjustment and trade liberalization efforts. The World Bank also released $350 million at the same time. But before that, Pakistani leaders had dexterously played on the fears of the Americans and even the Indian leaders about the extremely negative results of Pakistan's economic collapse. Various arguments were used, but basically three themes were recurrent: (1) in the face of severe economic difficulties, Pakistan might be tempted to share its sensitive nuclear technology and information with Islamic states or even rogue states like North Korea; (2) Pakistan was a moderate Islamic country; chaos in Pakistan could in fact affect neighboring Islamic states and the Middle East; (3) an economic breakdown in Pakistan could have a domino effect on India.

These factors, combined with the desire to "reward" Pakistan for not testing first, prompted the U. S. government to help Pakistan in getting an IMF package reinstated. Of course, this decision was subtly linked to U. S. nonproliferation benchmarks. The Indian government, in fact, complained that President Clinton's exercise of waiver on sanctions in November 1998, shortly before Prime Minister Nawaz Sharif's visit to Washington, was tilted in favor of Pakistan. The economic bailout was further buttressed when the United States returned to Pakistan $466.97 million for the undelivered F-16 aircraft.

In March 1999, the Republicans in the U. S. Congress moved legislation to wrest the initiative on removal of sanctions from the administration. Senator Sam Brownback, backed by the powerful chairman of the Senate Foreign Relations Committee, Jesse Helms, introduced legislation that would provide a five-year suspension on sanctions, mainly economic ones, with a possibility that the Pressler amendment could be repealed. Under the Brownback amendments of 1988 and 1999 the president was given the power to waive most of the provisions of these amendments.

The very signal from the Republicans that waiver on sanctions could go through without India and Pakistan signing the CTBT led Pakistan to the conclusion that there could be no negative economic consequences.

In April 1999, Pakistan once again sought to take high moral ground by flight-testing Ghauri 2 and Shaheen missiles in response to the Indian test of Agni. India was blamed for accelerating the missile race, and Pakistan presented itself as a respondent, not an initiator. The White House and the State Department only expressed regrets and concerns and did not impose any new sanctions. Another factor favoring India, and indirectly Pakistan, has been that corporate America and U. S. businesses in general have opposed economic sanctions. The U. S. Engage, a group of U. S. businesses, has spearheaded a campaign to convince senators and members of Congress to roll back all unilateral sanctions or, at least, to introduce some measure of moderation in them.

Clinton waived sanctions against India. Pakistan got waivers only for agricultural commodities and U. S. loans while the Pressler amendment remained in place. Under the Glenn amendment, both countries were barred from buying weapons from the United States, military financing, and dual-use technology. The economic situation remains precarious: General Pervez Musharraf's government claims the foreign currency reserves are at $1.6 billion. Pakistan is still heavily dependent on World Bank, IMF, and London and Paris Consortium loans and faces the challenging task of financing a gap of more than $5 billion in its balance of payments. The breathing space provided by the international creditors is set to expire by December 2000. Analysts predict there are very clear and definite limits beyond which Pakistan cannot increase its military outlay.

Army and Defense Decision making

In the past, the army was more or less the sole arbiter of Pakistan's defense requirements, and civilian leaders were comfortable with arrangements whereby the army alone dealt with these matters. During the Sharif government, this practice seemed to be changing because of two developments: overt incorporation of nuclear weapons into Pakistan's defense doctrine, and the authoritarian direction of Prime Minister Sharif 's government, whereby he diminished the powers and independence of major state institutions.

The decade after the death of General Mohammed Zia-ul Haq in 1988 was the longest time that the military had been out of direct rule since Pakistan's independence in 1947, though it remained the pivotal power behind the curtains. From backstage, it supervised the ouster of four governments since 1985 through forced resignations of prime ministers or dissolution of parliament. However, the proof of the army's reluctance for a takeover became evident on several occasions. Most recently in June 1997, when the government was at loggerheads with the judiciary and the president, the military could easily have climbed back into the saddle. Instead, from behind the scenes the military arbitrated the stalemate in the government's favor. In 1997, after the presidential powers to remove the prime minister were limited, the army had little option but to stage a naked coup without the facade of constitutionality that they had maintained in earlier coups.

In his statement to the UN General Assembly in September 1998, Sharif announced Pakistan's willingness to "adhere to" the CTBT. A day before Sharif was supposed to address the assembly, Chief of Army Staff General Jehangir Karamat had issued a statement that Pakistan should not sign the CTBT. It was reported that the prime minister changed his speech at the last minute and desisted from making a categorical announcement on the CTBT. Not appeased by the prime minister's backtracking, the army chief issued another stricture against government policies. He expressed dissatisfaction over the unilateral decision making process and proposed establishing a National Security Council; both statements were considered to indicate the military's mistrust of the Sharif government. During the events that followed, General Karamat had to resign.

Musharraf's sacking on October 12, 1999, a year after he had replaced General Karamat, proved to be the last straw for the army. It reacted by overthrowing the Sharif government. As chief executive, Musharraf formulated an interim government without a time frame for restoration of democracy.

In Pakistan, successive civilian governments until 1998 were willing to leave nuclear matters to the armed forces. The white paper on higher defense organization commissioned by Zulfikar Ali Bhutto in the early 1970s was the first and last attempt made by the political leadership to introduce a more active civilian role into defense planning and decision making. In the mid-1970s,

Bhutto also formulated the central hierarchical system of the Joint Chiefs of Staff Committee (JCSC), with the military subservient to the political leadership. However, Bhutto's attempts to place civilians at the helm of defense decision making were doomed: General Zia-ul Haq established the supremacy of the army from 1977 to 1985.

The Ministry of Defense, the Defense Cabinet Committee (DCC), and the Defense Council have all been criticized for neither having an independent source of information nor providing useful inputs in decision making. The nuclear establishment in Pakistan faces the problem of lack of coordination. Some of its entities are controlled by the Ministry of Science and Technology and others by the Ministry of Defense. A few are independent financially and only liaise with the army. Reportedly, when Prime Minister Sharif consulted A. Q. Khan, who heads the Khan Research Laboratories (KRL) and Pakistan Atomic Energy Commission (PAEC), about the time required for Pakistan to prepare for a test, Khan said it would take a month; PAEC chairman Samar Mubarikmand said that they would need only two days'; notice and some enriched uranium to test. According to the Pakistani navy's director of research, these entities are "like small islands not connected to each other."

PAEC and KRL have both made contributions to Pakistan's nuclear program and claimed credit for the success of the nuclear tests. However, the overall coordination and supervision of Pakistan's nuclear strategic planning and execution may be done by the Combat Development Directorate (CDD) of the armed forces. The rudiments of a central command and control system may have started taking shape in Pakistan. Newspaper reports indicated that in a meeting held on May 5, 1999, nearly a year after Pakistan tested nuclear weapons, the JCSC was deliberating on the proposed structural framework of such a system. Its basic components would be (1) a national command authority comprising some ministers and three services chiefs to be headed by the prime minister, who will have the authority to order the use of nuclear weapons; (2) a development control governing body; (3) a strategic force command in the hands of the JCSC chief; and (4) a secretariat for all these commands.

Coordination and Miscalculations

Coordination does not seem to be one of the government's strong points. After the Indian tests, Sharif stated that no decision on a Pakistani test had been taken, but Foreign Minister Gohar A. Khan nearly simultaneously said that it was not a question of "if, but when" Pakistan would test. Having assured its public of a "befitting response, " Pakistan was still conducting negotiations with the United States on an "aid package. " Nearly a year after the tests, the foreign minister said there was "no possibility" of a war with India, whereas Pakistan's permanent representative to the UN tried to convince world opinion that there were serious chances of war with India. General Musharraf said there is "zero" percent chance of war with India, but a minister of the Azad Jammu and Kashmir government said that a serious danger of a nuclear war with India exists. So, the Pakistani government still must determine not only who the ultimate decision makers will be in the nuclear realm but also what kind of doctrine would best suit Pakistan's threat perceptions and capabilities.

There is a long history of miscalculations and misinterpretations by Indian and Pakistani leaders seeking to enhance their security through steps that send confusing signals. In 1987, large-scale Indian military exercises dangerously close to the Pakistani border prompted Pakistan to deploy its own military divisions in preparation for war, which was defused only at the last minute. A second such crisis unfolded in May 1990, when fighting in Kashmir almost triggered a nuclear exchange between India and Pakistan when both countries had acquired covert capabilities. The incident was defused by high-level U. S. intervention. South Asia has seen three full-fledged wars between India and Pakistan; an ongoing military conflict at the Indo-Pakistani border on the Siachen glacier that has lasted for sixteen years; and a so-called low-intensity conflict in Kashmir coupled with violence sponsored by intelligence agencies across the Indo-Pakistani border.

An incipient dialogue between India and Pakistan after the tests had helped to develop a blueprint for nuclear restraint, though little progress was made in truly showing restraint. An essential adjunct of this exercise was the Lahore process, agreed to by Prime Ministers Nawaz Sharif and Atal B. Vajpayee and hailed by the rest of the world. The objective of this process was to avoid a conventional war between India and Pakistan, which had the potential of escalating into a deadly nuclear conflict.

While the world and the political leaderships of India and Pakistan were constructing these underpinnings of deterrence, the armed forces of Pakistan conceived a "brilliant" plan to extract optimal benefit from Pakistan's newly acquired capability. In their hopes of internationalizing the Kashmir dispute, they seriously miscalculated the response of the international community.

Prior to the developments at Kargil, serious concerns were being expressed about the inadequacy or absence of nuclear doctrines in India and Pakistan. India had started a strategic defense review, but it was not clear when it would be completed. Pakistan had also begun internal consultations on these issues.

It was against this backdrop that the misadventure at Kargil was played out. As the fighting along the line of control (LOC) intensified, defense strategists began to grapple with the following questions: Would the current crisis ignite a wider conflict and possibly the use of nuclear weapons? What was the threshold of provocation on either side to warrant use of a nuclear weapon? Were any command and control systems in place in Pakistan, even in their most rudimentary form? Was nuclear threat a political hyperbole meant just to scare the adversary when, because of the real fears of devastation of the civilian population of both countries, the nuclear option would never be exercised?

There is plenty of evidence to suggest that Pakistan's armed forces engineered the Kargil plan to, inter alia, test the strength of their unwritten nuclear doctrine. They assumed that Pakistan's nuclear deterrent would hamstring India in making massive strikes against the guerrillas for fear of nuclear conflict. They were proved wrong. Humiliated at their intelligence failure, nervous and flustered Indian soldiers pounded at the guerillas, and the Indian army introduced air power for the first time. Pakistan did succeed in "internationalizing" the Kashmir issue, but its leadership discovered to its horror that this meant Pakistan's total diplomatic isolation. The world community did not seem to care too much about Kashmir or the right of self-determination. Their priority was to avert a deterioration of the situation toward a nuclear war. The United States led the Group of 8 (G-8) countries in delivering a stern message to Pakistan that it had to reverse infiltration. Pakistan was told to assimilate the new reality that nuclear states cannot attempt to alter borders through conventional wars. Pakistan first balked at the suggestion but had to cave in under enormous international pressure.

Sharif 's hastily arranged meeting with President Clinton on July 4, 1999, was aimed at strengthening his position vis-à-vis his recalcitrant armed forces and involving the Americans in mediation efforts on Kashmir. Despite these efforts, the denouement of the Kargil crisis does not guarantee nuclear stabilization in the region or avoidance of accidental confrontation.

"Minimum" Credible Deterrence

The concept of minimum deterrence has always been in currency in South Asia. After the tests of May 1998, this term came to be used more often, though it remains to be defined. It has also been argued that a precise definition cannot be developed because it is merely a "concept. " The more one tries to understand it, the more it defies applicability to South Asia. Lawrence Freedman defines minimum deterrence as "the possession of sufficient nuclear weapons to inflict grievous harm on the enemy in retaliation and no more. " Going by that definition, it is problematic to determine what would be "grievous harm" and "no more. " The concept of "no more" does not fit into the picture comfortably with the inclusion of China in India's security calculus. In terms of arsenal, if India wants to equip itself well enough to deal with China, then the quantity and quality of nuclear weaponry goes beyond South Asian requirements. Will Pakistan be able to match India's capabilities in this realm for long? Can it afford to sit back and relax now that it has acquired "the capability"? Can it define for itself where it should stop refinement of nuclear weapons and missiles? Does it have a doctrine whereby it can spell out (even to itself) its requirement in concrete terms?

In spite of indications that there are no clear answers to these questions, the usage of minimum deterrence serves to tacitly appease the international community that a nuclear race in earnest has not begun in South Asia. The United States has repeatedly solicited from India and Pakistan, both directly and indirectly, an elaboration of their ideas on minimum deterrence, but it has remained unsuccessful in obtaining a quantitative answer. The Indian foreign minister is quoted as saying: "Minimum credible deterrence is not a physical quantification which is finite or fixed and limited in time. The policy has certainly a physical shape when translated physically, but that physical shape is determined by the security requirements of the time and that certainly is not something that can be spoken out." Prime Minister Vajpayee categorically stated the following before the Indian parliament: "While our decision is to maintain the deployment of a deterrent which is both minimum and credible, I would like to reaffirm to this House that the government will not accept any restraints on the development of India's R&D capabilities.

The Pakistani ambassador to the UN criticized this approach as "practically open ended" and not even remotely defining anything concrete. He added: "Given Pakistan's lack of 'strategic depth,' any threat of India launching a decisive and lightning conventional strike due to its overwhelming preponderance in conventional weapons will force Pakistan to adopt a more pro-active nuclear posture." Conventional military imbalance is repeatedly cited as the motivation for Pakistan's search for a nuclear deterrent. The two reasons given by Pakistan are that 75 percent of India's conventional assets, and especially its strike formations, are deployed on Pakistan's borders; and that India will continue to gain a decisive edge in the conventional field, with its indigenous defense production and its continued procurement of weapons from a variety of sources, especially from Russia and France. As the asymmetry between India's and Pakistan's conventional capabilities increases, the role of nuclear weapons in Pakistan's security is likely to increase proportionally.

Pakistan's nuclear policy revolves around three basic tenets: (1) a nuclear threat warrants a nuclear response; (2) nuclear force will act as a force multiplier to balance the asymmetry in conventional forces; and (3) there should be a regional solution of nonproliferation issues. Soon after testing and restoring "strategic parity" where India had disrupted the "strategic balance," Pakistan delinked its nuclear policy from that of India. This delinking has been a rather dicey affair. Despite public pronouncements of delinking its decision to sign the CTBT from India's nuclear practice, Pakistani leaders have had to belabor the point that they would continue to "monitor" what the Indians were doing. Pakistan also emphasizes that if India were to resume nuclear testing, Pakistan would review its position, and in case it had "adhered to" the CTBT, it would invoke the supreme interests clause as provided under Article 9. This unmistakably relinks Pakistan's nuclear policy to Indian nuclear policy.

Clearly, Pakistan will be conducting more work in the future to strengthen its nuclear capability. One such area will be missiles, since Pakistan has only forty frontline aircraft, as opposed to India's 300. This gap in capabilities would increase Pakistan's reliance on its missiles as a substitute for the air force's conventional role, thereby requiring Pakistan to build more and better nuclear missiles. In addition, Indian air superiority would keep Pakistan under the constant fear of India's launching a preemptive surgical strike against its strategic assets. Finally, if India were to destroy Pakistani planes, which are more vulnerable to attack, Pakistan would be short of delivery vehicles for retaliation. Another area that needs more effort is miniaturization and bomb design, and the third area is production of fissile material at a hectic pace before Pakistan gets roped into the CTBT.

After the release of India's Draft Report of National Security Advisory Board on Indian Nuclear Doctrine in August 1999, Pakistan's three best known strategists suggested induction of mobile nuclear-tipped missiles into the Pakistani arsenal, a "high state of alert" (as India deployed its bombs), and dispersal of these nuclear weapons. Such deployment, however, will make the command and control system more decentralized and accident-prone, as commanders at the operational level must attempt, under the tremendous pressures and uncertainties of battle, to prevent a decapitating attack against their nuclear forces. Deployment of nuclear weapons would pose difficult resource and technical challenges for a country like Pakistan, which lacks the heavy infrastructure and expertise required. Construction of new facilities, establishment of secure communications channels, and the rigorous personnel selection, training, and organization needed to maintain high levels of nuclear system security, connectivity, and responsiveness under alert conditions of dispersal, deception, and concealment may prove to be a formidable task. Deployment will also require attention to the daunting problems of passive defense of cities and industry against a nuclear attack. In the most authoritative statement so far on coupling and deployment, General Musharraf claims that "one of the main issues is geographical separation of the warheads and the delivery system. When that is so, there is no button. It is only when you couple them that you are ready, the button is on. But that is not the case. When the time comes, yes, there will be a button. And that button will be with me, of course. But we have not got to that as yet."

Constructing Doctrines

Two factors must be kept in mind. First, the government of Pakistan has made no elaborate public statements on the issue of doctrines, though Pakistani leaders and officials have made some laconic remarks on the subject from time to time. In a press conference in Islamabad on May 30, 1998, Foreign Secretary Shamshad Ahmed said, "Our nuclear doctrine is premised on deterring aggression, whether conventional or nuclear." Second, the government has initiated a process for evolving the doctrines, but it remains veiled in secrecy. Pakistani officials have stated: "It is a fact that while India and Pakistan have developed nuclear weapons they are still in the process of evolving their respective nuclear doctrines." Thus, Pakistani decision makers have clearly established a linkage between the development of their doctrine and that of India. This linkage, however, will lead them down a tortuous and unpredictable path. Their hopes to "evolve a common strategic vocabulary" via the Lahore memorandum of understanding (February 1999), so that India and Pakistan would move toward "meaningful arms control measures," seems anomalous. India does not want to "accept any arms control fetters." Furthermore, Pakistan cannot have a substantive dialogue with India on minimum deterrents unless both countries are prepared to exchange information on capabilities in the nuclear and missile fields.

Discussion on confidence-building measures (CBMs) has been a regular feature of the dialogue between the governments of India and Pakistan. On paper, there exist at least seventeen official CBMs. The prime ministers of Pakistan and India have a direct hotline to communicate with each other. A dedicated communication link (DCL) between the Pakistani and Indian directors general of military operations (DGMOs) exists to avert an accidental war. Both countries have agreed on several measures to notify each other of any military exercises to maintain transparency. CBMs have also been affirmed in the Tashkent Declaration, Simla Accord, Agreement on the Non-Attack of Nuclear Facilities, and Joint Declaration on the Prohibition of Chemical Weapons. However, the history of these official CBMs is replete with betrayals on the part of both countries. On numerous occasions, both sides have cheated each other. Instead of restoring confidence, these measures have been used to take advantage of each other. Mutual trust has been conspicuously missing.

The most telling example of this relates to chemical weapons. In August 1992, Pakistan and India agreed to a joint declaration on the complete prohibition of chemical weapons. Pakistan alleged that later, at the time of joining the Chemical Weapons Convention (CWC), India declared having stocks as well as production facilities for the express purpose of dealing "with any situation arising out of possible use of chemical warfare against India. " Another example is the Agreement on the Prohibition of Attack Against Nuclear Installations and Facilities, signed by Pakistani prime minister Benazir Bhutto and her Indian counterpart, Prime Minister Rajiv Gandhi, in December 1988. Among other things, it required an annual exchange of lists detailing the location of all nuclear facilities in each country. However, when the lists were finally exchanged in 1992, each side reportedly left off one facility. Similar experiences have occurred with other official CBMs like nonintrusion of air space and non harassment of diplomatic personnel. For instance, a commitment to keep delivery systems non weaponized (as mentioned in the Lahore memorandum) has not met with any progress because neither side will agree to data exchanges at a nascent stage of missile production programs or revelation of missile characteristics and numbers. Verification might require data exchanges on missile inventories, their location, periodic radiation checks on missile nose cones, and investigation of their absence outside designated sites in situations of doubt. (Because some of these weapons are road mobile, if they are not in their location they may have been deployed.) Given the state of Indian-Pakistani relations, agreeing to all this certainly would be a tall order.

Against this backdrop, the Pakistani and Indian foreign secretaries, meeting for their first formal dialogue following the test in October 1998 in Islamabad, exchanged a new cluster of rather ambitious but largely mutually exclusive CBMs. Addressing a press conference on October 18, 1998, following the dialogue, Ahmed said that Pakistan had proposed CBMs "which will signify our commitment to the reduction of tension, nonuse of force and peaceful settlement of disputes. " The framework suggested by Pakistan included proposals for the conclusion of an agreement on nonuse of force (a non aggression pact), the prevention of a nuclear and ballistic missiles race, risk reduction mechanisms, the avoidance of nuclear conflict, the formalization of a moratorium on nuclear testing, the non induction of antiballistic missile (ABM) and submarine-launched ballistic missile (SLBM) systems, a nuclear doctrine of minimum deterrent capability, and the mutual and balanced reduction of forces and armaments. With regard to Jammu and Kashmir, Pakistan proposed some nine security-related and twenty-three human rights-related "interim" CBMs. India matched them by suggesting its own proposals on an agreement on prevention of nuclear war, including through accidental or unauthorized use of these weapons, extension of the Agreement on the Prohibition of Attack Against Nuclear Installations and Facilities, advance notification of ballistic missile tests, improved communication links for verification of data exchange, and implementation of existing agreements.

At the foreign secretaries'; meeting in October 1998, the two sides only exchanged their proposals. However, at the Lahore summit, the prime ministers of India and Pakistan agreed to "take immediate steps for reducing the risk of accidental or unauthorized use of nuclear weapons and discuss concepts and doctrines with a view to elaborating measures for confidence building in the nuclear and conventional fields, aimed at prevention of conflict. " Their foreign secretaries, on the same occasion, agreed to provide each other with advance notification of ballistic missile flight tests and undertook to conclude a bilateral agreement in this regard. However, these pious intentions were put to the test when the two countries tested their ballistic missiles in April 1999. Pakistan complained that India failed to notify it about the Agni 2 test on April 11, 1999.

Pakistanis said that they learned about the test after an inquiry made by their high commissioner in New Delhi. Although confirming their intention to test, India did not provide information about the time and the date, they alleged. This drove home the point that efforts by both sides were tentative and primarily designed to gain international respectability.

However, an enunciation of nuclear policies, objectives, and doctrines may be more achievable. Pakistani experts have urged the government to "make a clearer declaration, linking its doctrine to realistic and rational strategic objectives. An ambitious or ill-conceived decision on nuclear deterrence by Pakistan may push India into a more competitive nuclear stance. Over-reliance on nuclear deterrence by Pakistan could thus prove counter-productive, dangerous, and financially exorbitant. " Pakistan has declined to accept India's offer of an agreement on no first use and countered this proposal by suggesting negotiation on a comprehensive non aggression treaty encompassing both nuclear and missile fields. Some Indian analysts have taken it for granted that a nuclear first strike is a principal part of Pakistan's nuclear doctrine: "Unless a nation wishes to practice deterrence through solely a first-strike posture, as Pakistan is doing, it has to invest in second-strike assets. "

The director general of inter services public relations in Pakistan has ruled out any move to downsize the army but did clarify that various options were under consideration to engage the military in the uplift of the country. This statement leads one to speculate whether Pakistan is planning to resort to a nuclear war with India at the first serious provocation and may well want to reinforce its conventional capabilities to tackle such a situation. It is not clear how far Pakistan will have to be pushed to decide on a first nuclear strike. In the ultimate analysis, situational variables may make all the difference.

Pakistan's doctrine will also have to take into account whether its nuclear capabilities should be limited to preparation for first strike or whether it should develop second-strike capabilities. It is generally assumed that Pakistan's doctrine will inherently revolve around a first strike. A number of factors support that assumption. First, Pakistan's asymmetry with India in the conventional field makes a first strike an equalizer. That is why Pakistan has repeatedly rejected

India's proposal for an agreement on no first use. Second, the development of a first-strike capability is less cumbersome for Pakistan. Investment in retaliatory or second-strike capabilities requires intense planning and enormous resources, which Pakistan cannot afford. One may also assume that Pakistan's pattern of fighting wars may necessitate the targeting of population centers and cities rather than strategic and military facilities; Pakistan does not have the quality and quantity of nuclear weapons to deal with sophisticated targets and, therefore, may just opt for inflicting "grievous harm" against population centers. Big population centers wiped out by nuclear bombs constitute unacceptable damage.

Public Opinion

Pakistan's public opinion has been traditionally supportive of its nuclear policy. Following the Indian tests, an overwhelming majority of the public wanted Pakistan to test. A survey, sponsored by the Joan B. Kroc Institute of the University of Notre Dame, before India's detonation showed 61 percent of the Pakistani elite supporting Pakistan's official policy of keeping "the option open," whereas 32 percent favored acquisition of the overt nuclear weapons capability. As many as 73 percent cited another Indian test as sufficient justification for Pakistani weaponization. A Gallup poll published just a day before Pakistan took the nuclear plunge found that 70 percent of the urban population surveyed favored a nuclear test; only 30 percent advised restraint. More than 80 percent said there was a high or modest chance of war with India, and 64 percent believed that India would use its nuclear weapons in case of war.

Support for the nuclear weapons program remained after the tests. Nevertheless, the last few years have seen the formation of small pockets within the elite ruling class that are prepared to think "rationally" on these issues. There has been no consensus on a reactionary policy based on the Indian nuclear policy after the tests. Some Pakistani analysts have called for an opening up of the defense policymaking process so as to make it more responsive to public opinion. This trend has been accompanied by a new mood for deliberation on nuclear doctrines, command and control systems, permissive action links (PALs) to prevent the unauthorized launch of nuclear weapons, intelligence, and transparency. Movement in these areas is circumscribed by (1) the defense and foreign policy establishment's unwillingness to share sensitive information, (2) inadequate substantive work in these areas, (3) a shortage of expertise on defense and nuclear issues in Pakistan, and (4) a very narrow base of educated people interested in the substantive terms of the debate.

According to a Stimson Center report published before the detonations, there were nearly forty dialogue channels operating in South Asia, ranging from nonofficial forums that serve as testing grounds for new policy initiatives to people-to-people exchanges that enhance transborder links. However, when the debate whether Pakistan should go ahead with a tit-for-tat answer to India's nuclear detonation was at its peak, the group suggesting restraint was too small to be counted. Although it was intensely debated whether Pakistan should test or not, those opposing it were branded as traitors. An overwhelming majority called for detonation. The major chunk of the population the lower-class and lower-middle-class sectors in urban areas as well as the rural areas of Pakistan supported a nuclear race with India. This huge segment of the society, whose opinion was also influenced by the press and the government's pro-nuclear statements, clamored for Pakistan's retaliatory tests in May 1998.

Hardly any subscribers to the view that Pakistan should unilaterally give up its nuclear program reside in Pakistan. According to polls, a marginal 6 percent favor the renunciation of nuclear weapons. Opinion in Pakistan on the nuclear issue revolves around two mainstream schools of thought: one that supports full "bomb-for-a-bomb and missile-for-a-missile" nuclear competition with India and another that takes the middle ground. The second school advocates neither a dismantling of the nuclear program and renunciation of nuclear weapons nor a full-fledged nuclear competition with India. No matter which group held power democratic leaders or the military public thinking about the nuclear program would not undergo any drastic change. Politicians frustrated with the moderate elements in public opinion have even demanded that the public abide by their promise "to eat grass but make the bomb."

It has been suggested that at certain times Indo-Pakistani leaders tried to address the nuclear and Kashmir issues rationally, but public opinion was not ripe for compromise. Former prime ministers Sharif and Inder Kumar Gujral tried to bridge differences between their countries, but the domestic atmosphere in these two countries forced them to pause. The Sharif government's decision to open trade with India came under severe criticism. When the Pakistan-India People to People Forum organized a get-together of people on the Wagah border to mark the fiftieth anniversary of their freedom on the eve of August 14, 1997, the press made a big issue out of it. Jamaat -i- Islami issued warnings that people trying to participate in the function would do so "at the risk of their lives." Only a handful of people dared to turn up at the border. Similarly, when a seminar was organized to discuss the repercussions of Pakistan's detonation of the bomb, the participants, including respected academic and known pacifist Abdul Hameed Nayyar, were physically assaulted by right-wing journalists and workers from Shabab-i-Milli, the youth wing of Jamaat-iIslami. In contrast, Sharif and Vajpayee were better prepared to handle public opinion when they met in Lahore in February 1999, although after the summit, they had to peddle their particularistic interpretations of the declaration in Lahore to pacify their domestic constituencies.

Islamic Groups

Traditionally, religious parties like Jamaat-i-Islami and Jamiat Ulema-iIslam have never mustered enough votes to play a dominant role in parliamentary politics. But their influence as a pressure group is disproportionately significant. The last few years have seen a resurgence of extremist religious groups in Pakistan deriving strength from the rise of the Bharatiya Janata Party (BJP) in India, the Taliban movement in Afghanistan, and the sectarian Shia-Sunni strife fueled by Iran and Saudi Arabia. Their frontline role in the proxy wars in Afghanistan and Kashmir gave further impetus to their growth. For most of these groups, people like Ramzi Yousaf, the accused in the World Trade Center bombing; Aimal Kansi, who was sentenced to the death penalty for killings outside the Central Intelligence Agency (CIA) headquarters in Virginia; and more recently, Osama bin Laden, are crusaders, not criminals. Lashkar-eTayyaba, the military wing of the Ahle Hadith sect, is considered as "the most effective fighting unit" among these groups after Harkatul Ansar, a party declared by the U. S. State Department to be a terrorist group. Both these groups are involved in the Islamic jihad (holy war) extending from Afghanistan to Kashmir. Mujahideen (Islamic warriors) return from the front to report on their fighting in Kashmir and Afghanistan to the annual congregations, attended by hundreds of thousands of volunteers from as far away as Myanmar, the Philippines, Chechnya, and Eritrea.

Not only are religious parties more prominent in the country, but also the mainstream political parties are becoming increasingly religious in their orientation. Benazir Bhutto's Pakistan People's Party (PPP) has entered into a political alliance with religious parties. Nawaz Sharif tried to please religious elements by attempting to introduce an amendment to Pakistan's constitution that sought to implement the shari'sa (Islamic code of life). If this amendment had been passed by the Senate (it had already succeeded in the National Assembly), Pakistan would have more closely resembled theocratic states like Saudi Arabia and Iran than the "moderate" nation that it claims to be.

Almost all religious parties were happy at Sharif's ouster in October 1999 The major reason was his withdrawal from Kargil. Both the Lahore declaration and the Washington accord were resented by religious groups. General Musharraf was initially seen as the savior who would pursue a hawkish agenda against India. The inclusion of a religious scholar in the newly formed National Security Council was allegedly an attempt to appease the religious groups. But reports of Musharraf's "secular inclinations" and his liking for a Turkish model when he spent a few years in Turkey as a youth have already caused concern among religious parties. Jamat-i-Islami's Qazi Hussain Ahmad censured the induction of liberals from the IMF and World Bank in the cabinet. On the nuclear issue, religious parties are extremely hawkish, rallying against the government whenever it tries to go "soft" on issues like the signing of the CTBT.

U. S. Role

Between 1974 and May 1998, Pakistan advocated nuclear restraint in South Asia while at the same time developing its nuclear capabilities. Its proposals on a nuclear weapons free zone

(NWFZ) in South Asia, renunciation of acquisition or manufacture of nuclear weapons, simultaneous signing of the Nuclear Nonproliferation Treaty (NPT), and acceptance of International Atomic Energy Agency (IAEA) safeguards remained unanswered by India. The United States and other nuclear weapons states were lukewarm toward Pakistan's proposals. In the aftermath of the May 1998 tests, the United States acted as an ardent promoter of the four nonproliferation benchmarks for India and Pakistan, namely the CTBT, the fissile material cutoff treaty (FMCT), export controls, and a restraint regime. Taking its lead from the Americans, Pakistanis developed their own proposals on a restraint regime and presented them to India in October 1998 during the talks between the foreign secretaries.

Pakistan had proposed convening a five-nation conference to discuss conventional and nuclear restraint, but this proposal was not accepted by India. After the tests, India's dialogue with the United States on security and nonproliferation issues seemed to work to Pakistan's advantage. In a sense, the trilateral dialogue initiated in 1998 between the United States, India, and Pakistan was a revival of Pakistan's proposal for multilateral talks on the whole range of security and nonproliferation issues, which Pakistan would have liked the United States to discuss eight years before. This time, India did not veto such talks because their modified format, with the United States talking to India and Pakistan separately, gave India sufficient leeway.

The Pakistani government repeatedly asked the United States to remove all sanctions against Pakistan, particularly those imposed under the Pressler amendment. Larry Pressler is seen as a symbol of the discriminatory U. S. policy toward Pakistan on nonproliferation. However, Pakistan's debacle at Kargil severely damaged both U. S. support for Pakistan and the nonproliferation aims that were being pursued in South Asia by the United States. In addition, the military coup in October 1999, though not condemned by the United States, prevented "business as usual" with Pakistan. As of this writing, the Pressler amendment has not been repealed by the U. S. Congress, and sanctions against Pakistan are currently more extensive than sanctions on India, a situation further complicated by U. S. legislation against countries under military rule.

Thus the United States has a critical role to play. For Pakistan it would remain the primary interlocutor, not only in removing sanctions but in engaging India in a dialogue. U. S. involvement would not alter any element of the Indo-Pakistani dynamic. However, Pakistan hopes that the United States would use its leverage with India, which could be decreasing as analysts suggest, to help address the most difficult issues, including the issue of Kashmir. The United States would also work closely with Pakistan on the modalities and rules and regulations governing export controls of tangible and intangible technology. Pakistan has repeatedly stated that it would welcome any assistance from the United States in the area of command and control systems. Further, Pakistan would remain amenable and sensitive to U. S. criticism on such issues as the functioning of democracy, good governance, human rights, narcotics, and terrorism. Pakistan would also aggressively solicit U. S. foreign direct investment (FDI).

Summary

The question of whether this may also force an unacknowledged delinkage of Pakistan's nuclear policy from that of India remains to be answered. Despite promises to "upgrade" its capability to ensure "survivability and credibility, " the new Pakistani foreign minister has also categorically denied engaging in "any nuclear competition or arms race. " India continues to say that its concerns are not Pakistan-specific and that India's wider considerations about its own security, particularly regarding China, will have to be addressed for any viable dispensation of arms control and a restraint regime. Pakistan insists that its program is India-specific. This response also conditions Pakistan's behavior and responses, yet an arms race per se may not necessarily be the outcome. Pakistan's development of its own nuclear capabilities need not necessarily follow rigidly the tit-for-tat (or knee-jerk) formula that it has been accused of using. Having consecutively responded to India's developments in the nuclear and missile fields, especially since May 1998, Pakistan has already established the veracity of its "respondent" plea.

Even if Pakistan's economy were to reach its optimal level in a fairly short period, the gap between India's and Pakistan's nuclear weapons capabilities can continue to widen. Deployment of nuclear weapons may be an area where Pakistan is put under considerable pressure to closely monitor and match India, which will be of crucial importance in high-tension and near-conflict situations. Nevertheless, a restraint regime, though in Pakistan's interest, is unlikely to succeed with its South Asian parameters. Pakistan would like to put U. S. weight behind its own effort to persuade India to agree to a strategic restraint regime. It would also like the United States to use its clout with India to convince the latter to address outstanding unresolved issues, particularly the issue of Jammu and Kashmir.

General Musharraf has not only shown a keen interest in the Indo Pakistani dialogue but has given no indications of massive investments in an elaborate nuclear deterrent. Though Foreign Minister Abdul Sattar has admitted to Pakistan's intentions of periodic upgrades, Pakistan may prefer to be reticent on matters such as the exact threshold of provocation required for Pakistan to make a preemptive nuclear strike. Therefore, neither Kargil nor the Draft Report have altered Pakistan's nuclear policy. After the release of the report, Pakistan's complaints of the damage it had dealt the SRR in South Asia and nonproliferation in general seem to have been taken in stride by the international community, which is fast adjusting to a nuclear South Asia.

Even after India's definition of its rather elaborate "minimum credible nuclear deterrent" in the draft, Pakistan seems to prefer to take time to deliberate upon its policies and keep its pronouncements ambiguous and noncommittal. Reticence could serve Pakistan's interests as it hesitates over precipitously declaring its assets or divulging information that could compromise its "supreme national interests." It should also give Pakistan sufficient room for maneuver and flexibility. However, Pakistan cannot escape from the inevitable responsibility of doing its homework on developing a nuclear doctrine and streamlining command and control systems, which according to General Musharraf has been taken care of to quite an extent. The exercise of internal consultations and discussions in this regard will have to be brought to a conclusion, but Pakistan will not be obliged to share with India or the rest of the world its conclusions on Pakistani strategic thinking until a higher degree of transparency vis-à-vis India becomes a reality. And while Pakistani decision makers wait for India to provide more information, they can harmonize their own positions, improve coordination between various agencies responsible for nuclear and missile management, and strengthen the institutional framework dealing with them.

In the meantime, emerging trends of extremism in Indian and Pakistani politics need to be watched closely. The ascendancy of extremist forces in either country could heighten volatility in South Asia and make the nuclear situation highly fissiparous. A developing, prosperous Pakistani society will be less prone to espousing and fortifying an environment of chronic hostility. In this context, the role of U. S. FDI and regional cooperation between India and Pakistan in economic and commercial fields will be of crucial importance. A sustained dialogue, launched imaginatively by Vajpayee and Sharif, may offer prospects for peace and development in the region, as Pakistan delicately balances its reactionary syndrome vis-a -vis India.

11

India's Space Program: Technology for Development

Introduction

India's space program is involved in two major activities it builds satellites used for remote sensing, meteorology, and telecommunications and, more significantly, builds rockets that can launch these satellites. These space technologies have found both economic and military applications. In the 1970s and 1980s, India's first satellite launch vehicle (the SLV 3) could only launch a light payload (typically, a scientific satellite) to 300- to 450kilometer-altitude low-earth orbits. Such lightweight low-orbit satellites did not have major military or commercial applications. However, SLV 3 was still powerful enough to be used as an intermediate-range ballistic missile (IRBM), the Agni missile. In the 1990s, India built far more powerful launch vehicles that can launch heavier satellites to the necessary higher-altitude orbits. These more advanced satellites have primarily economic applications but also offer potential military spin-offs. The polar satellite launch vehicle (PSLV) enables the Indian Space Research Organization (ISRO) to place its medium-weight remote-sensing (or reconnaissance) satellites into the optimal 800- to 900-kilometer-altitude orbit. The PSLV's successor, the geosynchronous satellite launch vehicle (GSLV), would permit ISRO to launch heavyweight communications satellites to the necessary 36,000-kilometer high-altitude geostationary earth orbit (GEO).

The capabilities of its space assets, and examine their political, economic, and geostrategic aspects. In the three decades since the inception of the space program in the 1960s, its technical objectives have remained fairly constant. However, under changing political and economic circumstances, India's space assets have been put to varying applications they have been utilized for socioeconomic development or to fulfill political objectives and have produced military spin-offs and commercial applications. India's space assets are now sufficiently advanced to find applications for power projection and force multiplication, which is achieved through satellite reconnaissance, intelligence, and communications. These increasing capabilities of India's space assets, combined with the modernization of India's conventional forces and the development of its nuclear forces, could significantly influence the strategic relationship between India on the one hand and Pakistan and China on the other. Such a geostrategic capability will have important security implications for the Asia-Pacific region.

History of India 's space Program

India's space program has passed through two stages of development. The first stage began in the 1960s and involved setting up an administrative framework and gaining experience with elementary rocket operations. ISRO was formed in 1969, and the Indian Department of Space was established in 1972. ISRO 's first chairperson, Vikram Sarabhai, persuaded India's political leaders to increase funding for the Indian space program at a time of national financial constraints in the late 1960s. Sarabhai was also instrumental in planning the gradual evolutionary development of the Indian space program. In 1970, he noted that "in a ten year time frame, one must acquire the capability not only of building telecommunication satellites such as INSAT-1 but also of launching them into synchronous orbit. To begin with, we have set as a goal the development of a satellite launcher of the Scout type once the basic systems have been developed for this, and experience acquired in operating them, it is estimated that a five year period from 1975 to

1980 should be adequate for the second stage of development of larger boosters. " The Indian space program has lagged more than a decade behind Sarabhai 's original schedule.

The latter phase of the first stage of India's space program focused on mainly experimental, low-capability projects that allowed Indian scientists to gain experience in the construction and operation of satellites and launch vehicles. In this phase, ISRO indigenously built (with foreign assistance) the Bhaskara earth observation satellites as well as a communications satellite (the Ariane Passenger Payload Experiment [APPLE] satellite) and conducted four flight tests of its SLV 3 satellite launch vehicle between 1979 and 1983.

The second stage of India's space program, commencing in the mid1980s, focused on-more capable mission-specific systems. It involved building the PSLV and its successor, the GSLV, which were designed to launch the indigenously developed Indian remote-sensing (IRS) satellite and a meteorology and telecommunications "Indian national satellite" (INSAT), respectively. With the PSLV commencing operational launches in 1997 after three demonstration launches and the GSLV project nearing its first flight test in the year 2000, India's space program is now emerging out of its developing stages and stands poised to join the ranks of the world's five advanced space agencies that have GEO capability.

India 's Satellite Launch Vehicle Program

Efficiency and economy of scale have been achieved in ISRO 's launch vehicles as each utilized some of the guidance and rocket propulsion systems from the prior vehicle. For example, for rocket propulsion, ISRO utilized a single 9-ton solid-fuel system in the SLV 3, allowing it to launch a 35-kilogram payload to a 300-kilometeraltitude low earth orbit (LEO). Three such systems were utilized on the augmented satellite launch vehicle (ASLV), enabling it to place a 100kilogram payload in a 450-kilometer-altitude LEO. The same 9-ton solid fuel rockets are also used in the Agni missile and as peripheral strap-on boosters in the PSLV.

The PSLV and GSLV involve two major propulsion systems a 130ton solid-fuel system, and a 37-ton liquid-fuel engine (whose design is based on the Viking engine used in the European Space Agency's Ariane launch vehicle), both of which are indigenously built. They allow the PSLV to launch a 1,000- to 1,200-kilogram payload (the IRS satellites) to an 800- to 900-kilometer-altitude polar orbit. The same systems are used on the GSLV and are supplemented with a 12-ton Russian-supplied cryogenic engine, which enables the GSLV to carry a heavier 2,500-kilogram payload (an INSAT 2–class satellite) to a higher 36,000-kilometer GEO. The first batch of GSLVs will use Russian cryogenic engines, but later GSLVs will utilize indigenously built cryogenic engines. ISRO is developing a 12-ton (7.5-ton-thrust) cryogenic engine for launching 2,500-kilogram INSAT-type satellites. It also plans to build 12- and 16-ton-thrust cryogenic engines for more advanced GSLVs that would place 3,500-kilogram and 4,000-kilogram communications satellites in GEO. These heavier satellites would carry a larger number of transponders and facilitate a greater volume of communications.

India's satellite launch vehicle program, though maturing, has not escaped failures – only five out of eight (62 percent) SLV 3 and ASLV launches were successful, and four out of five (80 percent) of PSLV launches have been successful. The first SLV 3 launch (August 1979), the first (March 1987) and second (July 1988) ASLV launches, and the first PSLV launch (September 1993) failed; the fourth PSLV launch (September 1997) reached the required altitude but did not precisely place the satellite into the correct orbit. In most of these cases, the hardware and major propulsion systems performed well; the failures occurred because of software problems such as the delayed ignition of stages, disturbances caused during the separation of stages, and the inability of the guidance and control systems to correct for trajectory deviations resulting from these disturbances. [5]

India 's Missile Program

India's missile program partly overlaps and is partly distinct from its space program. In July 1983, New Delhi officially embarked on an integrated guided missile development program (IGMDP). The IGMDP was initially allocated a budget of Rs380 crore ($130 million) and by 1994 had received over Rs780 crore ($275 million). The IGMDP was pursued through India's Defence Research and Development Laboratory (DRDL) at Hyderabad in South India, which is part of a larger military research and development apparatus, the Defence Research and Development Organization

(DRDO). The IGMDP aimed at building five missiles an antitank missile, two surface to air missiles, and the 150- to 250-kilometer-range Prithvi missile and the 1,500- to 2,500-kilometer-range Agni missile. The latter two systems form the backbone of India's missile programs.

The Prithvi missile has a propulsion system derived from the SA 2 missile of 1960s vintage. Prithvi missile production has been undertaken by Bharat Dynamics Limited, and sixteen Prithvi tests were conducted between 1988 and 1998. It bears little resemblance to ISRO's sounding rockets, which theoretically could have been converted for use as shortrange ballistic missiles. Thus, in terms of short-range missiles, ISRO and DRDL have pursued different rocket programs.

The 1,500- to 2,000-kilometer-range Agni missile is more directly derived from the Indian space program's SLV 3. In its early years, when it was headed by Vikram Sarabhai and Satish Dhawan, ISRO consistently opposed any military application for its dual-use projects such as the SLV 3. Eventually, however, the DRDL-based missile program borrowed human resources and technology from ISRO. In a move resented by the ISRO community, missile scientist Abdul Kalam resigned as head of the SLV 3 project at ISRO and moved to DRDL to direct India's missile program. The extent of human resource transfer between ISRO and DRDL should not be exaggerated barely a dozen scientists accompanied Kalam from ISRO to DRDL. Kalam designed the Agni missile using the SLV 3's first stage for propulsion, although the Agni's guidance system is different from that of the SLV 3.

The Agni 1 missile, which uses a solid-fuel first stage and liquid-fuel (Prithvi-missile-derived) second stage, was flight-tested three times between 1989 and 1994 to a maximum range of 1,400 kilometers. The project was then suspended in 1996; an official report noted that since all the objectives of the Agni technology demonstration project had been met, the project was being terminated. It added that the decision to develop and produce a missile system based on Agni technology could be taken at an appropriate time consistent with the prevailing threat perception. Following the Bharatiya Janata Party (BJP) government's assumption of office in 1998, India's nuclear and missile activities gained momentum. India conducted a series of nuclear tests in May 1998. In April 1999, India tested the rail-mobile Agni 2 missile to a range of 2,000 kilometers. This missile had a solid-fuel second stage built by ISRO.

MTCR Technology Embargoes

The missile technology control regime (MTCR) is a key international instrument aiming to halt the spread of ballistic missiles by denying states the necessary foreign technical assistance for missile or rocket development. India's space program had initially benefited from foreign technology transfers. For example, in the early 1970s, cooperation between ISRO and Germany's space agency resulted in the transfer of technologies relevant to propulsion and guidance systems, flight controls, material science, and wind tunnels. In the 1960s, foreign sounding rockets were launched from the Thumba range in South India, and the SLV 3 design was based on the design of the U. S. Scout rocket of 1960s vintage. However, despite foreign assistance becoming restricted with the advent of the MTCR, India's space activities were not halted.

ISRO and Indian industry indigenously developed a number of space and missile technologies after being denied their import these included shell catalysts for rocket fuel, magnesium plates for the Prithvi, radiation hardened integrated circuits for satellites, and maraging steel for rocket motor casings. This indigenous construction typically took five years for each system and resulted in approximately a 10 percent increase in expenditures. These obstacles were significant in the short term, but because ISRO was not operating under rigid time constraints, the slow pace of indigenous production did not hurt ISRO projects. The most publicized case of U. S. sanctions against ISRO those invoked in 1992–1994 against ISRO and a Russian firm, Glavkosmos, for the supply of cryogenic engines to ISRO has, in the middle term, not affected ISRO activity. ISRO eventually acquired cryogenic engines from Russia (albeit without the technology). Moreover, following the U. S. sanctions, ISRO embarked upon an indigenous program to develop cryogenic engines. The slow pace of this indigenous effort does not affect the GSLV project because Indian built cryogenic engines are not required for the first batch of GSLVs, which will use imported cryogenic engines. To summarize, the MTCR has delayed and increased the costs of ISRO projects, but it has not actually halted any project.

Indian Satellite Program

India's satellite program has achieved quicker and somewhat greater success than the satellite launch vehicle program. The first Indian satellites Aryabhatta, Bhaskara, and Rohini enabled Indian scientists to gain experience in satellite operations and were stepping-stones leading toward the development of more capable satellites. The 1,100-kilogram INSAT 1 satellites were somewhat unique because they had multiple roles of communications and meteorology, each carried out by different sensors on the same satellite. The INSAT 1 series was constructed in the United States by Ford Aerospace. The 1,900- to 2,100-kilogram INSAT 2 satellites are indigenously built and have transponders that are more powerful and more numerous than those on INSAT 1. Low-power C-band transponders typically support a few hundred phone lines each, whereas higher-power transponders in the extended C, S, and Ku bands each support a television channel or high-volume data and mobile communications. INSAT 2 meteorological sensors provide seven-minute repetitive coverage (useful for tracking and relaying back information on rapidly changing weather conditions such as cyclones), an improvement over the thirty minute INSAT 1 time frame.

INSAT satellites have been well utilized in the setting up of a national telecommunications infrastructure for example, INSAT 1B facilitated the extension of television coverage to over 75 percent of India's population, and subsequent INSATs brought almost all of Indian territory under television coverage. It should also be clarified that the theoretical promise of technological advancement translating into immediate socioeconomic benefits has not been realized. This is because much of India's population remains illiterate, despite being within the range of educational television programs. However, the INSAT series has found several more modest commercial and developmental applications. The INSAT 2 satellites provide telephone links to remote areas; data transmission for organizations such as the National Stock Exchange; mobile satellite service communication systems for private operators, the Indian railways, trucks, and road transportation; and broadcast satellite services used by India's state-owned television agency and by commercial television channels. The GSAT series are smaller satellites to be launched on the initial GSLVs.

From late 1999 onward, the INSAT 3 series began to replace INSAT 2. INSAT 3B was launched in late 1999 to provide fixed as well as mobile satellite services. This satellite would be followed by INSAT 3A, a multipurpose communication and meteorological satellite, which will also carry a meteorological data relay transponder and a search and rescue transponder. INSAT 3C would be a communications satellite similar to INSAT 3A, whereas INSAT 3D would be primarily configured as a meteorological satellite.

The IRS program has removed India's dependence on satellite images from Landsat and SPOT satellites. IRS 1A was launched in March 1988, IRS 1B in August 1991, and IRS 1C in December 1995, on Russian launch vehicles. IRS 1D (September 1997) and the IRS P series are launched on India's PSLV. IRS 1A and IRS 1B sensors had a resolution of 72 multispectral (in the visible and near-infrared band) and 36 multispectral panchromatic. IRS 1C and IRS 1D cameras have a better resolution of 23 meters multispectral and 6 meters panchromatic. The IRS P series are quick-turnaround low-cost versions of the IRS 1 satellites, costing approximately $15 million (far less than the $50-million IRS 1 series). IRS P3 carried a wide field sensor (WiFS) with 180-meter resolution to monitor vegetation. IRS P4 (Oceansat) is an ocean satellite launched on PSLV C2. IRS P5 (Cartosat 1) will provide 2-meter panchromatic images, and Cartosat 2 (to be launched in 2002) will have 1-meter resolution. IRS P6 (Resourcesat) will have 10-meter resolution sensors for detailed vegetation monitoring, allowing multiple-crop and species-level discrimination. IRS P7 (Oceansat) will have fishing applications, and IRS P8 and IRS P9 are intended for atmospheric and environmental studies.

The IRS satellites form the nucleus of the Indian natural resource management program. Regional remote-sensing service centers (RRSSCs) in five Indian cities and remote-sensing application centers in twenty Indian states have used IRS images for a wide range of applications. These include monitoring the environment in general, analyzing soil erosion and the impact of soil conservation measures, managing forests, determining land cover for wildlife sanctuaries, delineating ground water potential zones, mapping areas of flood inundation, monitoring droughts, estimating crop acreage and deriving agricultural production estimates, monitoring fisheries, surveying metal and mineral deposits and other mining and geological applications, and urban planning.

Political Economic Influences and objectives

When it was first conceived in the 1960s, India's space program was intended to play a significant role in a broader national policy of planned socioeconomic development. In general, technology held great promise of allowing countries to leapfrog over traditional stages of development and make a quick transition to an industrial and postindustrial society. Therefore, satellite communications, educational television programs, meteorology, and natural resource survey and management were and continue to be priority areas for the Indian space program. It has also been guided by strong political motivations: it was intended to symbolize India's high-technology achievements and thereby enhance India's prestige internationally, especially among the nonaligned group of nations. In addition, the Indian space program caters to a domestic constituency it retains considerable popularity with the Indian public, and successful satellite deployments and launches function as national morale boosters.

India's space program was sharply upgraded in mid-1972, following the 1971 war with Pakistan and New Delhi's perceptions of a de facto U. S.-Pakistan-China convergence against India in that war. ISRO was separated from the Atomic Energy Commission and placed under the Indian Department of Space, which was set up in 1972 and assigned rigid schedules for meeting technical and applied objectives. Soon after the Indian nuclear test in May 1974, ISRO 's then-director, Satish Dhawan, stated in July 1974 before an Indian parliamentary committee that India had the ability to produce medium-range missiles with locally developed solid fuels and guidance systems he was referring to the SLV 3, which was actually still some five years from its first flight test.

In the mid-1980s, the most significant military spin-off from India's space program emerged when the SLV 3's first stage was used in the Agni missile. The PSLV and its successor, the GSLV, however, are more likely to find commercial rather than military applications. The commercial aspect of India's space program became prominent in the 1990s, when, influenced by India's economic liberalization program, ISRO's goals widened to include offering its space services on the international market. The Antrix corporation was set up in 1992 to serve as the marketing arm of India's space agency.

One analyst noted that ultimately, "India's intentions may be judged simply from the level of technological capabilities, whether for development or defense purposes. Does the technology achieved in the civilian nuclear and space program provide India with credible nuclear weapons and delivery systems? The answer to this question is yes if not now, then at least in the near future. " India's space program has already provided an intermediate-range nuclear delivery system the Agni missile. In addition, India's satellites and satellite launch vehicles have other potential military and strategic applications, which are discussed below.

Nuclear Weapons Delivery Systems

India's first satellite launch vehicle the SLV 3 was modified into the 2,500-kilometer-range (with a 1-ton payload) Agni IRBM; India's subsequent satellite launch vehicles, the ASLV and PSLV, despite being more powerful than the SLV 3, are unlikely to find applications as ballistic missiles. The Agni is relatively light and therefore transportable; it provides for an ideal second-strike mobile IRBM system. The Agni gives India a minimum but not a secure deterrent against China, on account of its range limitations. China's most important cities, Beijing and Shanghai, are 2,500 kilometers away from (insecure) launch sites in northeastern India, and therefore the 1,500-kilometer-range Agni 1 or 2,000-kilometer-range Agni 2 cannot reach these targets. The ASLV and PSLV theoretically have a 4,000-kilometer and 8,000-kilometer range, respectively, and would bring China's heartland within range when launched from secure launch sites in eastern and central India. In practice, however, it would be difficult to deploy the ASLV and PSLV as mobile, truck- or rail-mounted missiles, on account of their strap-on boosters (which greatly increase the effective diameter of the system and make it hard to deploy on mobile launch units). Moreover, the PSLV is extremely heavy and very inefficient in terms of weight-to-payload ratio; a PSLV-derived missile would weigh some 170 tons, whereas most ICBMs of capabilities similar to the PSLV weigh under 100 tons. Therefore, both the ASLV and PSLV, if they were to be used as nuclear weapons delivery systems, would have to be deployed in silos, the construction of which would be easily detected. Clearly, Indian political leaders need to give more detailed consideration to the use of the ASLV and PSLV as nuclear weapons delivery systems.

Reconnaissance

The IRS satellites are primarily used for civilian applications, and the present IRS system only offers a moderate military reconnaissance capability, with the drawbacks of poor resolution and limited frequency of coverage. The LISS-3 (linear image self-scanning) cameras on IRS 1C have a 23-meter resolution in the visible and near-infrared band, permitting the detection of large military installations. The panchromatic (PAN) camera on IRS 1C has a resolution of 6 meters panchromatic, sufficient to broadly detect surface ships, aircraft, tank formations, and ballistic missile units but not to precisely identify these objects.

Two important shortcomings of the IRS satellites should be noted. First, the visible and near-infrared cameras do not permit viewing through cloud cover and dense foliage. Second, the IRS satellites are unable to provide round-the-clock monitoring. The 23-meter-resolution LISS-3 camera on IRS-1C has a twenty-four-day repeat cycle (i.e., it revisits a given location once every twenty-four days), whereas the 6-meter-resolution PAN camera has a forty-day repeat cycle. These shortcomings in repeat cycles are likely to persist even with the advent of new IRS satellites; for example, a combination of four such satellites, each with a twenty-four- to forty-day repeat cycle, would together provide at best a six- to ten-day repeat cycle. Thus the IRS system would be of little use in providing daily or hourly updates of a battlefield situation to military command centers.

These drawbacks can be overcome on an ad hoc basis by changing the IRS orbit to yield a shorter repeat cycle. ISRO could also increase the resolution from its satellites by simply moving them to a lower orbit. From their present 800-kilometer-altitude orbit, the LISS-3 and PAN sensors have 23-meter and 6-meter resolutions, respectively; when placed at a lower altitude (typically a 300-kilometer-altitude orbit), they would generate militarily useful 8-meter multi spectral and 2-meter panchromatic images, respectively. The above adjustments in satellite orbit would expend more fuel and decrease the satellite's life cycle.

New Delhi's defense bureaucracy has occasionally experimented with the use of IRS images for reconnaissance purposes; the Indian air force's airpower doctrine seeks to use space assets for surveillance and battle management. As discussed above, India's current satellites have limited military reconnaissance capabilities. The Indian military could, however, build a series of dedicated military reconnaissance satellites based upon existing IRS technology. This option becomes feasible because only small incremental costs would be involved in utilizing existing infrastructure from the IRS project and using a proven launch vehicle (the PSLV) to construct and deploy a dedicated military reconnaissance satellite system. The Kargil military operations in the summer of 1999 may have accelerated Indian efforts to use its PSLV-launched satellites for military reconnaissance, as India's military sought to use the forthcoming Cartosat 1 for such purposes.

Command, Control and Communications

India's communications satellites have some military capabilities but do not have specifically designated military functions. The INSATs can be used for multiple access digital data transmission, teleconferencing, and remote area emergency communications, features that would be well utilized in a command and control network and for search and rescue. The constraints on effective military use of the current INSATs stems first from their inappropriate frequency range. The C-, S-, and Ku-band transponders on INSAT 2 operate primarily in the ultrahigh and low frequency ranges, which is typical for civilian communication satellites; these frequencies can easily be jammed. Dedicated military communications satellites operate in ranges susceptible to jamming. A second limitation of the present INSAT systems is that the INSAT transponders have a limited capacity; furthermore, almost all INSAT transponders have already been leased for civilian and commercial use, leaving few available for India's military. The limitations of the present INSAT system for military communications can be corrected by releasing some transponders for military use. New Delhi could also opt to build dedicated communications satellites, a project that is technically and financially viable because costs are lowered by utilizing existing INSAT infrastructure. The PSLV can be used to launch 500-kilogram communications satellites (which would carry only a few transponders and facilitate a small volume of communication) into GEO, and the advent of the GSLV in this decade would give New Delhi a launch vehicle for heavier and more powerful 2,000-kilogram military communications satellites.

Strategic Utility of India's Space Assets

India's space assets provide it with first-generation military communications and reconnaissance capabilities, which are currently limited but can be considerably enhanced with upgrades to the present generation of systems. India's Draft Report of National Security Advisory Board on Indian Nuclear Doctrine of August 1999 specifically calls for space-based communications and reconnaissance systems. These systems would affect the calculations of regional security planners on both conventional and nuclear issues.

India's satellites are presently unable to maintain round-the-clock coverage of Pakistan's military activities; indeed, they failed to detect military intrusions in the Kargil sector of Kashmir in 1999. But they do permit the buildup (and regular updates) of a database of Pakistani targets for Indian security planners. If New Delhi acquires dedicated military reconnaissance satellites that provide daily coverage of Pakistan's military installations, this would give it a counterforce (military target–oriented) capability versus Pakistan. The fact that Islamabad's nuclear weapons delivery systems could be more easily destroyed by an Indian first strike would then considerably weaken Pakistan's deterrent and consequently make Islamabad more likely to consider a preemptive first strike of its own. These factors have important implications for regional security because they decrease the stability of deterrence in the subcontinent.

India's missile programs such as an advanced version of the Agni 2 or a future Agni 3 and military ASLV would provide counter value (city- directed) capabilities against China. Although it would take at least until 2003 for India to develop a modest nuclear deterrent force against China (comprising twenty to thirty Agni 2 missiles and more powerful Agni 3 missiles), India's satellite reconnaissance systems enable New Delhi to counter Chinese conventional threats in the short term. India's satellite intelligence capabilities give its military planners invaluable tactical and strategic information on Chinese military forces in Tibet. The IRS satellite system with its six-day repeat cycles or a future series of dedicated military satellites could give India's armed forces sufficient early warning about the movement of Chinese military forces from central China toward Tibet and India, thus facilitating the deployment of Indian conventional forces in time to counter any movement of Chinese troops.

With the maturation of India's satellite-based reconnaissance, intelligence, and communications capabilities, Pakistan's potential nuclear deterrent against India would become highly vulnerable to an Indian first strike. Further, New Delhi could also effectively counter a Chinese conventional military threat. Thus, the China-Pakistan strategic balance versus India would be radically altered. China would be unable to rely on Pakistan's nuclear deterrent versus India, and Pakistan would have less confidence in China's ability to balance India. The United States and Southeast Asian nations would also more seriously consider balancing any future emerging Chinese threat with India's limited nuclear deterrence and increasing satellite reconnaissance- and communications-backed conventional capabilities against China.

Commercial Applications

The PSLV and GSLV are likely to find commercial applications in the period 2000–2005, after a series of additional flight tests demonstrates their reliability to potential clients. Two small satellites, a 100-kilogram Korean satellite and a 50-kilogram German satellite, were launched (along with the heavier IRS P4) aboard PSLV C2 in May 1999. A 100-kilogram Belgian satellite would be similarly launched, along with an IRS satellite, on a future PSLV. ISRO did not gain much by launching these lightweight micro satellites, but it did acquire a market reputation that would help secure more lucrative heavier payload satellite launch contracts in the future. ISRO prices are comparable to those on the international market a

PSLV launch costs approximately $20–25 million, and a GSLV launch is expected to cost $70–80 million, and the anticipated global increase in demand for communication satellites would enable ISRO to secure at least a few commercial satellite launch contracts. Yet ISRO 's low launch rate would generate only a small annual revenue that would barely recover the considerable investment made in the PSLV and GSLV program. In general, because of its inability to be financially self-supporting, ISRO would have to rely on funding from India's government, and its activities are likely to be slowed in times of a national financial crisis.

The situation is somewhat more promising for India's satellites. IRS 1C images of 6-meter panchromatic resolution (better than those from their previous main competitor, 10-meter-resolution SPOT images) have found a number of clients, though ISRO imagery will eventually encounter competition from 1- to 3-meter-resolution imagery on the international market. The International Telecommunication Satellite Organisation (Intelsat) has signed a ten-year $100 million agreement with ISRO to lease eleven Cband transponders on INSAT 2E. The INSATs are also a major asset for India's state-owned television agency, which plans to use six INSAT 2C transponders to widen its broadcast footprint to the Middle East and Southeast Asia. Although the low capacity on the INSATs (each INSAT supports a smaller number of voice circuits or television channels than other communication satellites) makes them cost-inefficient, the extremely high demand for telecommunications in India and the Asia-Pacific region will ensure that INSAT capacity is fully utilized. ISRO plans to have approximately 130 transponders by 2002 (compared with sixty-seven transponders in 1998) and also plans more Ku-band transponders that allow a higher volume of communications than existing C-band transponders.

Costs, Manpower and Organizational Structure

India's space expenditures have increased gradually from an annual average of approximately $70 million from 1965 to 1980 to $85 million in 1979–1980, $230 million in 1985–1986, $300 million in 1995–1996, and $350–400 million in 1998–1999 and 1999–2000. ISRO employs more than 15,000 personnel (both scientific and technical), and in addition a large number of institutions more than

500 public and private sector firms, national laboratories, and academic institutions undertake research or build components for ISRO; some 50 percent of the ISRO budget is outsourced to these subcontractors. In recent years, approximately 45 percent of the ISRO budget has been utilized for operational satellites and 35 percent for satellite launchers, with the rest used for space applications at various regional space centers and administrative costs.

ISRO relies primarily on government funding and therefore remains under constant pressure to secure political support. India's political leaders, including the prime minister, are always invited to space launches in order to secure their endorsement for ISRO activities; they routinely oblige, especially because of the prestige associated with the Indian space program. At the same time, expenditures on science and technology endeavors are at times called into question. Thus, even following prestigious events such as the 1997 PSLV launch, one editorial in an Indian newspaper criticized the Indian space program because it had not created commercial spillovers. It noted that the technological feats of space launches were "not the sort of prowess that will make us a richer, healthier or more literate nation. India has to ask what its space program can deliver apart from grandeur.

The Indian space program fares poorly compared with the world's major space agencies, especially in terms of its annual launch rate. It also lags ten to fifteen years (one technological generation) behind the space programs of Japan, China, and Europe and about twenty years (two technological generations) behind the U. S. and Russian space programs. However, in terms of output per unit of expenditure, the Indian space program compares favorably with the space programs of other nations; it has provided modest socioeconomic, political, military, and commercial benefits without consuming a large portion of New Delhi's financial resources.

Summary

India's space assets were developed for economic purposes but have found some (and offer the potential for further) military applications. In purely economic terms, India's space infrastructure cost several hundred million dollars over a period of two decades from the 1970s through the 1980s, an infrastructure that did not provide immediate tangible economic or military benefits. It was only after this initial investment that India's more advanced space assets of the 1990s found modest economic and military applications. India could well have purchased such space services commercially on the international market at prices only marginally more expensive than indigenous services. Thus, in purely economic terms, India's space assets have cost more (when taking into account the sunk costs) than the revenues they have generated. However, locally supplied space services do not have to be paid in foreign currency, and this saving.

A second set of benefits from indigenously developed space assets is political in nature. First, space assets provide international prestige and have foreign policy spin-offs. For example, ISRO would be in a position to offer a larger volume of PSLV, IRS, and INSAT services to other states in the period 2000–2005, thereby reinforcing India's political and economic ties with these nations. Second, by acquiring technological autonomy over its space assets, India can use them not only for economic purposes but also for potential military spin-offs. The IRS-PSLV and INSAT-GSLV projects demonstrate ISRO 's ability not only to build but also to launch militarily useful and strategically significant reconnaissance and communications satellites. Doing so would greatly enhance India's power projection and force multiplication capability in the Indian Ocean and Asia-Pacific regions and thereby affect the strategic balance in these regions.

An increasing number of nations are likely to have national satellites, but ISRO is among of group of only eight space agencies that have a significant satellite launch capacity. The somewhat limited capabilities of India's space assets compared to those of the world's major space-faring agencies (especially when assessed in terms of quantity and volume of services and to a lesser extent in terms of quality) restrict their performance and international competitiveness but do not significantly detract from their modest socioeconomic, political, military, and commercial utility. Within a few years, the technological advancement of India's space assets could well coincide with and further facilitate India's emergence as the major economic and political power in the Indian Ocean region and as an important player in the Asia-Pacific.

12

Air and Sea as National Power of Security

Introduction

It is essential, nevertheless, if air power is to reach its full capacities and to become a tool of progress rather than retrogression and destruction, that we understand it and gauge its relationship to national power as a whole. This is so because air power viewed incorrectly will lead to the formulation of distorted national policies, and will bring ruin instead of progress to mankind. Although air power has yet to reach maturity, and any analysis must necessarily be incomplete, fortunately, our knowledge of sea power as an element of national power can be of great assistance in this effort.

National power is an amalgam of two principal elements: the inventory of a nation's resources and the national strategy which gives those resources direction. For example, it has become customary for the editors of American newspapers and magazines to print maps and charts on which population, raw materials, refineries, and military statistics are represented by rows of men and ships and guns or other appropriate symbols. These are supposed to be indices of strength; they are supposed to represent power, and in a certain graphic way they do. But power is more than stockpiles of metals or weapons or other static materials. In order for resources to be transformed into power, they must first be energized, must be given direction, and must either be put to use or prepared for use. This is why charts of national power which show only economic resources and military potential are incomplete. They cannot describe by symbols the cultural development of peoples or the character of governments, which in the last analysis determine the ends to which power will be put; nor can they describe the strategy by which those resources will be employed to achieve the desired ends.

It is not difficult, once national power is considered as a combination of resources and the strategy which develops and uses them, to see how sea power fits into the scheme of things. Sea power, being the ability to use the seas for one's own purposes, permits the nations which possess it to gather their own resources in peacetime and in war, and to exploit them; it enables them to draw upon the resources of friendly, allied, and neutral nations in time of conflict while denying them to the enemy; and it allows them to carry those resources, once they are turned into fighting goods, to whatever points along the oceanic pathways of the world that they are needed. Merchant fleets and naval ships, two indispensable elements of sea power, combine to make this possible. The principal reason why sea power is a critical element of national power is, therefore, that it enables the resources of the maritime nations of the world to be brought to bear upon the enemy. It provides the wherewithal to take inanimate resources from the ground, turn them into fighting goods, and put them into the hands of the soldier at the front.

This makes the connection between sea power and national power or its directing agency, strategy, crystal clear. Strategy is the art of the possible the logistically possible. It is the art of directing a nation's resources for the maintenance of national security. Strategy is inseparable from logistics because both are concerned with resources. One provides them; the other employs them, and strategy can do no more than logistics permits.

At the highest level of military decision, it is frequently discovered, when it is too late, that a nation's strategy is restricted to what is logistically feasible rather than to what is strategically desired. Military history is replete with instances of this kind - instances when strategic decisions were based on logistic capacities rather than on strategic theories, or when strategic decisions could not be implemented because of logistic failures. These failures, moreover, were due countless times to lack of shipping or failure to command the sea due, in short, to the breakdown or absence of some indispensable element of sea power. Thus, strategy in a global war, or in a war of any kind which must be carried on across the seas or nurtured by maritime trade for any considerable length of time, is heavily dependent upon what sea power can perform, that is, on what is logistically possible.

Air power so far has assumed a different shape. Its primary characteristic is mobility, which is generally considered to be the dynamic element of national power and an essential feature of successful military strategy. The basic attributes of air power's mobility are, however, speed rather than range, and maneuverability rather than carrying capacity. Therefore, air power does not have either the logistic or economic features which make sea power so vital a part of national power. Built around a weapons-carrying vehicle, its function has been primarily military rather than commercial, and the heavy emphasis which has been given to its unique ability to strike swiftly at the enemy's vitals from any compass point has tended to promote one of its military functions at the expense of others equally important.

Air power has tremendous potentialities, both economic and military, which are yet untapped. There are two reasons for this. In the first place, the airplane has not yet achieved mastery of its own medium; i.e., it is still too earthbound, still tied to bases on land and at sea, unable to operate in its own element except sporadically and intermittently. In the second place, we have not yet developed a comprehensive view of what air power is and what its potentials are. By confusing the airplane with air power and thinking of the former solely as a weapon, and primarily as an instrument of total war, we are stultifying the latter. By associating air power in our minds with one military service, and more especially with the Strategic Air Command, we have failed to recognize that it has other elements that are equally important -those which are based upon the sea, those which are vital as close support for ground forces, those which have to do with airlift and logistics, and those pertaining to air defense. Not until we begin to formulate a concept of air power in all its aspects, which includes all of its component parts, will air power attain its full development and its proper relationship to national power.

It was not unnatural when the airplane first made its appearance in World War I that it should be viewed simply as another weapon which, because it was able to soar above the battle lines, could serve most usefully as eyes for the military commander on the ground. Gradually, its missions were extended to operations against the enemy's troops, dumps, and supply lines, but the soldiers long continued to restrict it to these tasks and to consider it merely as an adjunct of the ground forces. Naval officers, by and large, were likewise intent upon employing the airplane solely to fulfill the traditional naval mission of controlling the seas. To most airmen, however, the plane was not merely another weapon; it was both a unique instrument of destruction and the agent of a new force in the world, air power, which would soon be as potent in military affairs as land and sea power had been, down through the ages. Unable to induce ground officers to accept this broader view of the airplane's potentialities, airmen began to fight for the establishment of a separate and independent service, which would be unhampered in developing its own doctrine and the air weapons which could make that doctrine a reality. Each service was not unnaturally guided by its own needs, and each developed its own particular philosophy of national security. The introduction of the air weapon into the arsenals of the world, therefore, almost immediately raised the question of whether air power should be conceived of as a separate and distinct element of national power, or simply as an adjunct of the two existing types of military power, land and sea.

Three distinct ways of thinking about air power have emerged from this controversy as the airplane has grown in speed, strength, and versatility, and as the weapons it can carry have mushroomed in destructive power. Each of these theories has certain merits; when they are combined, they provide a working basis for employing the airplane as it exists today. However, none is really broad enough to spur the development of aircraft into new fields, or to provide for a comprehensive philosophy of air power. Having been developed largely by military thinkers, these three theories continue to restrict air power to its military connotations. Air power, if it is to reach its maximum potential, must create an intellectual framework which will allow for and, indeed, will encourage research and development of aircraft which will be masters of their own element: immune to weather, capable of safe and sustained flight, and with greatly enlarged carrying capacity.

The first of these theories, having come down in a straight line from the older concept of the airplane as an adjunct of ground and sea forces, is surface oriented. Before Hiroshima, it was assumed that the ultimate decision in war depended upon destroying the enemy's surface forces, because as long as these were intact the people's will and capacity to continue fighting would remain unimpaired. The theory recognizes, of course, as any concept must today, that air power is vital to winning the surface battles, but, despite this fact, its advocates believe that armies and navies remain the dominant factors in war. Sea power remains essential for supplying both air and ground forces overseas, and for sustaining one's allies. Armies are essential for occupying the enemy's territory, particularly those points in which the locus of political power is concentrated. Surface forces are the primary instruments of both political and economic control, and the air weapon, while it is an incomparable tool of military action, cannot fulfill these essential functions.

The advocates of this theory, it is to be noted, are air minded, but they refuse to subordinate surface power to an air weapon. General Gavin recently expressed the view in this way: "Air power was the most significant entity of our war power in World War II. Without it, we never could have established ourselves on the Normandy Peninsula, and we might not have established ourselves in southern Europe. But air power was most effective as a component of our over-all war power, not as an independent entity." In corroboration of his views he quoted Walter Millis: "Strategic air power was perhaps the most striking feature of the Second War; it produced ruins more vast and somber than those of antiquity, of a kind which civilized man had never expected to see again. But if it was the machine's most gruesome and dramatic contribution to warfare, it was by no means militarily the most significant. The combination of the tank and the tactical airplane continued to dominate most theaters down to the end."

With this in mind, Gavin concludes that air power's greatest contribution will continue to be in support of surface forces; yet, since he believes that anti-aircraft defenses of the future will restrict the striking power of combat aircraft, he anticipates a greatly expanded role for the airplane in transport and logistic functions. "The true lesson of the airplane in World War II is," he says, "that, like other forms of mobility in the past, air mobility is most useful when employed to move man and his means of waging war to the area of decision, and there to continue to work closely with him."

Since the advent of atomic and hydrogen bombs, of course, the surface theory of air power has had to recognize that peoples and governments may be compelled to surrender, or may even be deterred from initiating a thermonuclear war by the mere existence of strategic air power, but it insists that the political objectives of war can no more be obtained today than they could before by means of strategic bombers. The future of mankind lies in exercising restraint in the use of air power, and this means that both sides must keep their objectives and their weapons at a level which will permit political objectives to be achieved. The strategic component of air power must therefore be held in reserve as a force in being. Only its tactical and logistic components should be employed in war, and these necessarily remain adjuncts of surface warfare. Air power must be exercised tactically only, i.e., in support of the ground forces which can achieve limited objectives, for strategic air power once unleashed will wreck the very national power it is intended to strengthen. It is a mistake of policy to place such heavy emphasis on an unusable element of air power at the expense of all its other elements, which are so vital to the survival of the surface forces and to the achievement of political objectives. For this school of thought about air power, in other words, limited war has become the only acceptable form of conflict.

At the opposite extreme, straight from the writings of Douhet, Mitchell, and Seversky, is a theory which views strategic air power as so decisive in itself that surface forces either become mere appendages or cease to exist at all. In the United States, its earliest and most brilliant exponent was Brig. Gen. William Mitchell, whose ideas grew rapidly with the development of the airplane and, indeed, whose concepts outpaced technology. For Mitchell, however, "One axiom, with its corollaries, never varied," as historians of the Air Force remind us. "To him the airplane was first and last an offensive weapon. In the vast theater of the air there were no frontiers, no battle lines, no terrain features." Thus, subject to the limitations of range, bases, and weather, the airplane could operate against any enemy objective. The only defense against an air force was therefore another air force, and, conversely, the first step in offensive operations must be the destruction of the enemy's air force. With this successfully accomplished, command of the air would be attained, and the only question then was to what end the resulting air power should be applied.

Mitchell early and correctly sensed that the great range and speed of future aircraft would modify the isolated position of the United States, and would render its cities subject to devastating attack. One aim of American air power should therefore be to counter this new threat. His solution, which does not vary by more than a hairsbreadth from that being followed today, was made public in Winged Defense in 1925. Our aircraft must henceforth strike directly at the enemy's "centers of production of all kinds, means of transportation, agricultural areas, ports and shipping." Powerful air attacks resolutely pressed home against the United States could never be completely stopped, but their effects could best be minimized by aggressive air action on our part. Accordingly, for the United States, war would probably become first a struggle for air bases, followed swiftly by a campaign of strategic bombard merit.

By 1930 Mitchell's air doctrine was complete. It began with a familiar military concept: "War is the attempt of one nation to impress its will on another nation by force. The attempt of one combatant, therefore, is so to control the vital centers of the other that it will be powerless to defend itself." Originally, armies and navies were used to defend these vital spots, and war was largely a matter of overthrowing the enemy's military forces, a problem which became increasingly difficult after the invention of the machine gun. The airplane, Mitchell asserted, changed all that. "The advent of air power which can go straight to the vital centers and entirely neutralize or destroy them has put a completely new complexion on the old system of war. It is now realized that the hostile main army in the field is a false objective and the real objectives are the vital centers. The old theory that victory meant the destruction of the hostile main army, is untenable." If anything, Mitchell had even less respect for sea forces. He cavalierly dismissed the aircraft carrier as an instrument of naval power, and, as early as 1925, insisted that the

United States" could entirely dispense with her sea-going trade if she had to, and continue to exist and defend herself." This being the case, he said, it would be far better for the nation to put its military appropriations and its energies into an active offensive weapon designed directly to defeat the enemy, instead of dissipating its own strength in an indecisive type of warfare.

Despite the astonishing changes wrought in aircraft and weaponry in the two succeeding decades, Alexander de Seversky, writing in 1950, had little to add to what Mitchell had already expounded. Where the relatively short range of planes in the 1920's required ground and naval forces to secure and support overseas air bases, Seversky maintained that the planes of his day did not, and air power therefore could be divorced entirely from surface forces. Moreover, like Douhet before him, Seversky was certain that by eliminating the burden of ground and naval arms, air forces could be given such overwhelming strength as to assure complete air supremacy and total victory in any future war. It would be strategically unsound to divide the nation's war-making potential three ways in order to support an "outworn triphibious method" of waging war, when by concentrating on air power we could rule the skies.

Few people today are willing to state the air supremacy theory so baldly, and yet, most of its advocates find themselves forced to approximate Seversky 's conclusions by the hard facts of current Soviet air strength. In 1948, the Report of the President's Air Policy Commission headed by Thomas K. Finletter put the case this way: "Our national security must be redefined in relation to the facts of modern war. Our security includes, as always, winning any war we may get into; but now it includes more than that. It includes not losing the first campaign of the war and not having our cities destroyed and our population decimated in the process of our winning the first campaign." The Commission reluctantly came to the conclusion, therefore, that such relative security as could be achieved in the nuclear air age was to be found only in being strong enough either to deter attack, or to smash it at the earliest possible moment. This would be difficult but not impossible, the Report continued. "We can be supreme in the air by the weight of our air power." In order to accomplish this, however, our "Military

Establishment must be built around the air arm. Of course an adequate Navy and Ground Force must be maintained. But it is the Air Force and naval aviation on which we must mainly rely. Our military security must be based on air power."

Basically, this has been the military policy since adopted by the United States, and yet, almost a decade after the Finletter Report, testimony before the Symington Committee indicated that if this view of air power were to be followed, our policy must veer ever closer to the Seversky thesis that surface forces must be reduced in order to attain air supremacy. The Russians, it was said, had built up their long-range air force to a point where it already surpassed ours, and would soon be a match for the combined strength of the medium- and long-range bombers of our Strategic Air Command. The testimony of Gen. Curtis LeMay, as summarized in the Committee findings, was that "it is not sufficient for us merely to match the Russians in strategic airpower. To be safe, we must have strategic airpower of sufficient strength to absorb any surprise attack and, even after suffering the heavy damage incident to such an attack, be able to retaliate with an effectiveness that would assure victory."

It was "axiomatic" to Gen. Carl A. Spaatz that the more retaliatory planes we had, the more chance we had to deter attack, and to destroy the enemy if deterrence failed. Moreover, testimony seemed to indicate that it was the long-range bomber, rather than the medium-range plane based overseas, on which LeMay pinned his hopes. Finally, it was evident that some Air Force spokesmen place little, if any, faith in the strategic striking power of naval aviation. Such forces would not only be committed initially against naval targets, but they would probably not be in a position to strike at the vital strategic targets which lie far inside the Soviet Union. "A nation's air-atomic strength," the Committee stated, "is not measured simply by the number of bombs in its stockpile. Rather such strength should be measured primarily by the ability to deliver [them] in the shortest possible time." It was therefore clear from the testimony that the strategic airmen were presenting the thesis that air power means long-range, home-based bombers alone, and that these should be built in such numbers that there will be little left in the way of military appropriations for other types of force, including not only land and sea but air as well. The current advocates of air supremacy thus have been compelled to accept the full implications of the Seversky thesis by their fear of the growing strength of the Soviet long-range air arm, and by their conviction that an all-out thermonuclear war is the most likely and most dangerous form of conflict that confronts us.

The third theory of air power extant, if it can be dignified by calling it a theory at all, is a kind of composite of the first two, moderate in tone, taking what it considers the best of each and eliminating the more extreme features. It is exceedingly difficult to express in definite and concrete terms because it is still seeking a solution to the frightful dilemma created by the marriage of the airplane and the nuclear bomb, and therefore is still somewhat nebulous and incoherent. In essence, this theory holds that although air power is by all odds the most telling kind of military power, there are definite limits to what it can accomplish. It is decisive as a deterrent to and instrument of total war; along with land and sea power it is essential for the implementation of foreign policy; and in local wars it is an indispensable aid to surface forces. Certainly today a good many people accept these premises, even if they have not attempted to combine and consolidate them into a systematized theory of air power. Nevertheless, a theory is implicit in these beliefs, and its source can be traced back to concepts which were being expounded at the Air Corps Tactical School in the 1930's.

Technological developments have, of course, altered or modified these old concepts to a considerable extent, but it is evident that in seeking a philosophy of air warfare, the Air Corps produced a set of ideas that can be taken as the basis of today's theory, and they are worth examining for this reason. In brief, the Air Corps Tactical School was preaching a doctrine which may be summarized as follows. The objective of war is to break the enemy's will to resist, and, since this requires offensive action or the threat thereof for its accomplishment, all components of the armed services must be prepared to exert the maximum pressure on the enemy. The immediate missions of the armed forces as a whole may be four in number: "defeat of the enemy's army, navy, or air force; the occupation of his homeland; pressure against his national economy; or operations directed against vital centers within his country." These missions can be executed most effectively by cooperation among the three military services: air, ground, and naval. This is so because each has limitations and abilities peculiar to itself. Aviation, for example, "can aid ground forces to gain territory, but only ground forces can occupy an enemy's land." Similarly, "aviation alone cannot protect our merchant marine or our troop movements by sea, but it can unaided accomplish the other functions of sea power." The latter statement indicates the extent of Mitchell's influence on the Air Corps of that day, but the important part of the concept is that "of the three arms, only aviation can contribute significantly to all of the designated missions."

The primary mission of the air arm, according to this doctrine of the 1930's, is to strike at the whole of the "enemy's national structure." This is so because "under conditions of modern warfare the military, political, economic, and social aspects of a nation's life are closely and absolutely interdependent, so that dislocations in any one will bring sympathetic disturbances" in all. Economy of effort, of course, makes it necessary for each arm to attack "that phase against which its weapons are most potent." But a nation may be defeated simply by interrupting the "delicate balance" of its economic structure and especially of its "industrial web," and these are vulnerable directly to the air arm alone. The Air Corps' primary responsibility is, therefore, to destroy the enemy's industrial strength, in order to disrupt the other essential aspects of his national life.

The similarity between this concept of the 1930's and the present moderate theory of air power is apparent. Each gives air power the dominant role in war but recognizes its limitations. However, strategic nuclear air power in the hands of the Russians has now modified the older concept in two important respects. First, it compels modern doctrine to emphasize the "threat," or deterrent aspect, of offensive air power and to shift initial air strikes from the enemy's industrial centers to his air and missile bases. And second, nuclear energy has forced today s moderate air theorists to deemphasize the strategic aspect of air power as unsuited to the nuclear age, and to give greater weight instead to the surface arms.

This latter change is most strikingly illustrated in the writings of Sir John Slessor, who more than anyone else has attempted to put the moderates' views on paper. In the Pollak Memorial Lecture which he delivered at Harvard University in 1954, and which was subsequently published, he remarked candidly: "I have spent many years of a rather long military career in trying to impress upon other people the truth that Sir Winston Churchill stated in his speech at Boston in 1949 that 'for good or ill, air mastery is today the supreme expression of military power.' Nowadays I find myself more often concerned to emphasize that air power cannot do everything. I believe that air power, allied with atomic and thermonuclear energy, has had the result that total war as we have known it in our generation, has abolished itself incidentally the only way it could be abolished."

The possible elimination of total war as an instrument of policy does not, however, abolish wars of all kinds, as Slessor was quick to point out. The Communists have not deviated from their aim of world domination with the coming of nuclear power; they have merely shifted from the tactic of frontal assault to that of indirection. They may attempt to restrict their program of undermining the free world to political and economic maneuvers, by pursuing the "tactics of the termite" as Slessor phrased it, but it is to be expected that the tensions of the cold war will break out occasionally into open military actions, such as in Korea and Indo-China. "Against this sort of local aggression," says Slessor, "t he Air Force can be an indispensably valuable partner, but it cannot in itself be decisive. Air power, in the full sense of the term, is essentially an unlimited instrument."

It is precisely at this point that Slessor and a good many other air theorists go awry. Too many of them tend to make air power synonymous with the air weapon, and thus distort the meaning of air power. Slessor, by linking air power with its most destructive and dramatic weapon, the strategic bomber, leaves himself no choice but to consider it "an unlimited instrument" of war. Air power viewed in this light is an unlimited instrument, but this is its narrowest meaning the one accepted by most air theorists today. In its "full" sense air power means more than that. It is the ability to command the air, so that land and sea forces may exercise limited control in their own right; indeed, it means command of the air for a whole variety of purposes, including air defense, reconnaissance, airlift and transport, and attack upon enemy surface and subsurface forces.

The organic air components of the Army and the Navy are as much a part of air power as the tactical and strategic air commands of the Air Force and, under certain circumstances, may be just as important. They not only give armies a mobility never known before, and add incomparably to the striking power and defensive capacities of navies, but they give to both a range of vision which permits these surface forces to employ their power with a swiftness and accuracy never before achieved. By removing the blindfold from the surface forces and thus adding to their mobility, ground and naval air arms have increased by a thousand fold the offensive and defensive strength of the forces which they serve. It is the failure to give full recognition to the importance of these features and functions of air power, thus reducing it to the level of an air weapon pure and simple, that has precluded us from viewing air power in its most comprehensive sense.

A correct assessment of air power requires consideration of a number of factors not generally taken into account. In the first place, while all forms of open and regular warfare are now dominated from the air, the more unorthodox forms of conflict are not. Air power properly conceived and executed may be of assistance against guerillas and partisan forces, but it remains ancillary to operations on the ground. Air power, moreover, may prevent the outbreak of major war and, in conjunction with surface forces, may prove to be a deterrent to smaller ones as well. It can have a major role in limiting the spread of conflict by transporting ground forces swiftly to the scene of conflict, and thus help to keep the struggle between freedom and slavery within non-military bounds. Neither air nor sea forces have, however, the ability to occupy territory, or to exercise the discreet and discriminating control over the political forces of a nation that ground forces have. The day has yet to arrive when air power can dominate and control the processes of cold war, for the affairs of men are still deeply rooted in and controlled from the ground. The cold war has indeed reaffirmed what history has long demonstrated, that the forces which can actually occupy a country "have a political significance out of proportion to their military strength." NATO trip-wire forces in Western Europe and Soviet armies in Central Europe testify to this fact. The Russians, it is to be noted, did not quell the Hungarian Revolution by the use of air power; they returned to Budapest with tanks and men.

Air power has, however, tremendous political and economic potentialities for the cold war which we must be prepared to exploit, as soon as the airplane becomes a true vehicle of transportation. If control of political processes is not yet a major attribute of air power, we should not forget the part it played in breaking the Communist blockade of Berlin. The airlift, it has been claimed, "was our first political victory of the cold war." If, in its infancy, air transport could sustain a city for a year and a half by airlifting more than 4,000 tons of food and fuel and other necessities of life daily, it holds out possibilities which only the air power fanatics of an earlier day dared to conceive. Mitchell was one of these. A full generation ago he pointed out that "transportation is the essence of civilization." And it is for us on the verge of applying nuclear propulsion to the airplane to consider what its implications for air power will be.

It is, of course, impossible to predict the effects of nuclear propulsion on air power with any degree of precision, nor is it the place of the historian to attempt it. He can only suggest that air power may take on some of the historic characteristics of sea power, and that in the role of navies past and present, there may be much of relevance for analysts of air power.

It is necessary to make one preliminary observation in order to avoid misunderstanding. Sea power has normally had one important characteristic which air power does not currently share. Command of the sea, unlike command of the air, has given the nations which possessed it control of economic processes of vast importance. For this reason, sea power is generally considered to have two principal aspects: military and economic, or more correctly, naval and commercial. Perhaps the best historical example of these two aspects is to be found in the exercise of British sea power during the Napoleonic Wars. Sea power furnished England with two weapons: naval superiority and money. The first enabled her to supply, reinforce, evacuate, and support by amphibious operations her own armies on the Continent and those of her allies; the second enabled her to procure cash to support by subsidies the enemies of Napoleon on the Continent, and provide the lifeblood for her own war effort. Moreover, command of the sea not only protected England from invasion and sustained her and her allies, but so constricted France that Napoleon was driven to measures that ultimately ruined him.

It is extremely unlikely that air power will ever possess this economic attribute to any such far-reaching extent. Certainly it does not today, as a comparison of sea and air freight of 100,000 long tons of supplies per month across the Pacific, from San Francisco to Australia, graphically demonstrates. It would require fewer than fifty surface ships but more than 10,000 cargo planes to haul this freight. In terms of manpower, the sea route would call for about 3,200 men, including gun crews in time of war, while air freight would require more than 120,000 highly trained men for flight crews alone. The sea route would expend some 165,000 barrels of fuel; the air would eat up almost nine million barrels. Moreover, while no tankers would be required to keep the freighters fueled, eighty-five large-sized tankers would be needed to fuel the planes, thus necessitating command of the sea in any case.

With four-fifths of the world's international commerce moving across ocean routes, and with air freight able to handle only a fraction of that commerce, it is not to be wondered that sea power has commercial advantages unknown to air power. Supply by sea is the primary lifeline of dozens of nations. For example, the great majority of nations with which the United States has allied itself in the struggle against Communism are maritime states located along the periphery of Eurasia and in the Western Hemisphere. Being dependent upon each other in peace and in war for the raw materials and finished goods which spell prosperity and strength, they are heavily dependent on coastal and overseas trade, and this is a dependence upon the sea which air transport at the moment cannot measurably reduce. Indeed, air transport will probably never replace sea transport as an economically feasible commercial proposition.

The military aspects of air power, on the other hand, have revolutionary prospects for the future, which we must now begin preparing to exploit. The military possibilities and emergency transport potentialities of bulk-cargo nuclear-powered airplanes seem to indicate that air power will have startling new roles in the future of hot war and cold war alike. It may well be that, while the performance of ships and aircraft carriers will be increased to a limited extent by nuclear propulsion, the airplane will be well high revolutionized, and air power along with it.

It is in the effort to gain some conception of the future role of air power as a military instrument that an examination of sea power becomes most useful. If in doing so we visualize the air power of the future as having greatly increased transport capacities, the analysis can be doubly meaningful. It might even be fruitful to note how often the term "air power" can be validly substituted for "sea power" in the discussion that follows. In any event, it will be necessary to bear in mind how heavily naval power is dependent upon its air arm, because in fact, much that is attributed to sea power today is in truth air-sea power, or the power of aircraft based at sea.

The number of tasks which sea power must perform in order to maintain the security of the free world is staggering. Indeed, the role of sea power is currently in the ascendant, exercising as it does new and more critical functions, and it should be considered one of the most important factors governing national strategy today. In order to give this statement credibility, it must be realized that while national strategy must be formulated in terms of the capacities of the nation, it must also take account of the nation's weaknesses and needs. On both of these counts sea power is a factor of critical importance. Not only is it our most distinct asset our greatest advantage over the Communist world but it demonstrates our dependence upon continuing freedom of the seas. In framing national strategy, then, we must at once exploit this advantage and decrease our vulnerability on the oceans, by developing naval power of great strength and versatility.

Sea power is our greatest asset for a number of reasons. Obviously, we can claim no advantage over the Communist bloc in terms of concentrated manpower. On the other hand, the enemy's advantage in ground forces can, to a considerable extent, be neutralized by the judicious exploitation of our command of the seas, because one of the main characteristics of sea power is its ability to land and to support sizable contingents of troops, at almost any point on the globe which is accessible to open water. By surprise operations on the periphery of the Communist dominated Eurasian land mass, the enemy can be compelled to hold large numbers of troops from the front lines as a strategic reserve. This seems to be the effect which the Korean War had for three years upon the Red Chinese, in their bid to take over Indo-China.

In terms of air power, it may be assumed that we still have the edge over the Soviet Union in some categories certainly in sea based airplanes and mid-range bombers but it cannot be assumed that this margin of superiority will be maintained forever. The strategic air component may be equalized, when missiles come of age, or may be neutralized by the fact that each side will be loath to employ weapons of unlimited destructiveness. On the other hand, if thermonuclear weapons should be used in the immediate future, it may be the mobility of sea power, or more correctly sea based air power, which will enable us to remain supreme in this field. Moreover, without sea power all of the planes of the United States would have to take off from their home bases, because we could not maintain overseas bases. This would place the Soviet Union at least on a par with us. Indeed, without sea power, there is reason to believe that Russia would have an advantage in the air war, for the heavily industrialized and urbanized society of the West is probably more vulnerable to bombing than is that of the Communist orbit. Thus, it is through sea power that the odds against us in ground forces and land-based power are reduced, and perhaps even tilted in our favor.

The advantages which this asset gives us can best be measured by considering, first, the critical needs of the free world maritime confederation. Our vulnerability is based upon several factors. In the first place, no nation is completely self-sufficient, but insular nations and those whose prosperity depends primarily on maritime trade are particularly dependent upon continuing freedom of the seas. It cannot be said that the Soviet Union is completely self sufficient, but it is much more so than we are, by reason of its central position in the Eurasian land mass. It is evident from the Russians' naval structure that they have correctly assessed the implications of these contrasting positions. Based as it is, primarily on submarines, plus a few cruisers, a large number of destroyers, and a shore-based naval air arm, the mission of the Soviet Navy is one of denial and defense rather than control of the seas. The Soviet submarine force, now approaching the 500 mark, is designed to rip the maritime nations asunder, by denying them the use of the high seas, and their fleet is intended to prevent the United States from projecting its power, via the sea, into Eurasia. By agreements with Red China for the use of warm-water ports in the Pacific, the Russians have gained greater opportunities to support raiding operations and countermeasures than their own geographic position permits. In other words, although the Russians have no need to use the high seas themselves, they are preparing to deny them to us and to defend against the offensive operations which we may have occasion to launch in behalf of our overseas allies.

It is important for us to realize how astutely the Russians have probed the second major weakness of our position, and have sought to exploit it by their naval strategy. They have recognized that the distribution of international power is no longer spread among a number of naval powers, which, if necessary, could fight it out on the high seas, but is spread between one nation and its satellites, which dominate the land masses of the globe, and another with its allies, which for all practical purposes have a stranglehold on the sea surfaces of the world. In other words, we have something akin to a bipolar world split along the axis of the sea and the land. Thus, on the one hand, we have a group of nations primarily dependent upon land transportation and, on the other hand, another group almost wholly dependent upon sea communications, and these two groups coincide with the power distribution of the world. If it should ever come to a prolonged war between these groups, the continental powers would have to strike their enemies upon the sea in order to win, while the naval states would be forced to project their power inland in order to achieve results. Each side would be most vulnerable inside its own orbit one, the land; the other, the sea and the battles would be joined along the rimlands and coastlines of the central land mass where land and water meet.

If the maritime alliance can win these battles, and keep the land powers cooped up in their own sphere, they will not only ensure the safety of their vital commerce and defend their allies on the periphery of Eurasia, but they will be in a position to extend their power inland. If they should lose these engagements, however, and permit their enemies access to the seas, their own commerce will be subject to destruction, their overseas allies will be lost, and defeat will only be a matter of time. The Communist bloc is held together to a considerable extent by military power, the suppression of the Hungarian revolt being a singularly brutal example of this. The free world is sustained by economic ties. The Communists, then, are bound to be sensitive to military action which could endanger one of the pillars of their internal strength; we are open to severe risks at sea, because the life blood of our economy flows through the great ocean arteries.

It is for this reason that sea power has become more important to the free world, rather than less; it is the reason why national strategy must be governed so largely by the requirements of sea power, as well as by what sea power can accomplish in a military struggle. We need each other to control the seas, so that we can maintain our commerce and sustain each other economically in peace and war, and so that we can hold the rimlands in order to protect our allies from the physical and psychological battering which neighboring Communist countries can give them.

No consideration of the needs of sea power would be complete without some reference to the elements of sea power, which are generally understood to be geographic positions, bases, merchant and naval shipping, and industrial capacity. These are the essentials of sea power, and, where they do not exist, sea power collapses. It is the task of strategy, then, to see that these are maintained. Strategy, of course, cannot change the physical geography of the world, but it can do its best to see that those nations whose geography is adapted to sea power are brought within one's own orbit or, conversely, kept out of the enemy's. It is no coincidence that all of the NATO nations are maritime states, or that the United States has sought close ties with maritime nations along the periphery of Eurasia and in the Western Hemisphere. National strategy also calls for the creation of bases appropriate to sea power; it calls for the growth of industries among the non-Communist nations. It is enough to note that the United States has tried to fulfill these demands by negotiating base agreements and by programs of military and economic assistance.

So much for the needs of sea power the need to hold what we have, to maintain control of the seas, to contain Communism within its present bounds so that the maritime nations will not be overwhelmed. What about its capacities? National strategy cannot be content merely to look at the negative side of things; it must discover what sea power can do in a positive way, and gear itself to those capacities.

By all odds, the greatest advantage of sea power is its flexibility and versatility, and this makes it of supreme importance as a factor in national power. National power, it is to be recalled, is an amalgam of resources and the strategy which directs them. Strategy is concerned with allocation of resources, and the first thing that the policy-maker must determine is allocation for what for what objectives, for what purposes in peace, and for what kinds of wars the future may bring. The types of war we expect to fight will play a vital part in determining the kinds and proportions of land, sea, and air power we develop, but, since the future cannot be predicted with certainty, the problem of strategy is to be as versatile as possible. Only in this way can we defend against all kinds of contingencies, and exploit all kinds of opportunities, and it is sea power which can best provide this flexibility today.

In a catalogue of the advantages of sea power, the contributions which naval power can make in war are particularly important. Naval power, to be sure, is only one element of sea power, the military element, but it provides the cutting edge and is perhaps an appropriate place to begin.

Traditionally, naval power has been able to make itself felt decisively only when wars have extended over considerable periods of time. This does not mean that naval power will count for nothing in a quick war; an earlier discussion of the role of naval and subsurface forces in total war makes this clear. Moreover, the addition of air power, as an organic component of naval force, speeds up the whole tempo of operations at sea in wars of more conventional variety. It is in these lesser conflicts that naval power will probably make its most vital contributions in the future. The amphibious operation, for example, remains a potent instrument in conventional war, even though its feasibility in nuclear war remains untested. The ability of the Navy to keep the enemy off balance by making swift descents upon his coast, or preferably through the coasts of his satellites and our allies, therefore, must be considered an advantage until events of the future prove otherwise. The territorial extent of the Soviet orbit is vast; the distances from its production centers to its borders is in some places overwhelming; its road network is sparse, and the terrain sometimes rugged. A comparison of the carrying capacity of the Trans Siberian railroad on the one hand, and the Pacific Ocean on the other should be cause for concern to the Russians, if they really hope to keep their Far Eastern naval bases operating.

With restless and uneasy satellites to be kept under military occupation, with loyal Communist neighbors to be supplied and reinforced, with naval and air bases on the rimland to be guarded, and with numerous threats of amphibious landings along thousands of miles of frontier, our sea power must be giving the Communists much to think about. If they have any true conception of the potentialities of sea power, and most certainly they do have, they will think carefully before tying themselves up in a large-scale land offensive against any of our Eurasian allies. As for launching an invasion across the sea, or into any area outside their own land mass, they know this is an impossibility as long as we control the oceanic pathways to aggression. Even in peacetime, they have found it impossible to foment Communist revolution, and make it stick, in lands which cannot be supplied from Russia by overland transportation.

The list of functions successfully performed by naval forces in the Korean War is an impressive demonstration of the role that naval power can play in conflicts along the Sino-Soviet rimland. Naval forces conducted amphibious assaults against defended shores, as well as ferrying operations to the shores still in friendly hands. They delivered both naval bombardment and air attacks, ranging from close support of advancing and retreating troops to deep bombardment missions, in order to disrupt enemy communications and channels of command. They participated in blockade and patrol operations and in mine-sweeping of coastal and adjacent waters. Naval forces not only supplied and reinforced United Nations troops, but evacuated them to fight again elsewhere. Sea transport was fully capable of bringing United Nations forces from all over the globe, and supplying them against land powers whose bases were in their own backyard.

When these accomplishments are added to the achievements of naval power in the last world war, we can only conclude that in conventional wars of any kind, great or small, where the opponent can be gotten at directly or indirectly from the sea, naval power is an essential of victory, and, if adequately conceived and developed, it will undoubtedly play a meaningful role in nuclear wars as well.

Turning from naval power to sea power in its broad context, it is already evident from earlier references that control of the seas is essential for the maintenance of our position in the cold war, and to keep its tensions from bursting into uncontrolled warfare. It is the primary means of keeping the oceanic confederation strong and prosperous, and if there is any military instrument which can raise and lower pressure in a controlled way, that instrument is sea power.

Specific instances of the role of sea power in the cold war are numerous. Without greatly stretching the imagination, or giving undue credit to sea power, it must be admitted that it played a part in helping to rejuvenate Western Europe through the Marshall Plan, and it procured and distributed the means to sustain Greece and Turkey in their fight to retain independence. It forestalled an imminent threat of invasion by Red China against Taiwan. It evacuated 300,000 refugees from North Vietnam when the Communists moved in there, and, more recently, it evacuated several thousand United States citizens from the Suez fighting zone.

It is important to note that while sea power performed some of these missions in the midst of war, much that it accomplished was done by peaceful means. The Taiwan illustration is evidence of the fact that sea power represents one force that does not have to invade a country in order to make its influence felt. From the western Pacific to the Mediterranean, we have had ample evidence that sea power enables a nation to move forces from one point on the globe to another, and to deploy them legally and effectively without aggressing. Sea power has permitted the nations of the free world to concentrate their influence and power on the front line, along the borders of the Communist world, to contain the forces which constitute such a threat to our way of life, and it has done this fairly economically in terms of blood, if not of treasure.

It should be evident from this brief survey of sea power that its greatest advantage is in support of our policy of containment along the rimlands of Eurasia. It is equally evident that air power must be fitted into this concept, if it is to play a major and continuing role in the struggle against Communism. Air power, to date, has been looked upon too readily as the means by which the rimlands can be disregarded, for it is expected that, in total war, the two great heartland powers will trade air punches to the point of mutual exhaustion. Air power is, of course, already contributing heavily to the catalogue of naval assets just listed, but it must be enabled to do more than that. What has to be done now is to consider how much more significant the role of air power will be when it comes to master its own element, and the air begins to serve as a true medium of transport. It may be that the struggle for the rimlands of Eurasia, which the United States has been engaged in for a decade, will then become really weighted in our favor for the first time. This is reason enough to make us begin thinking now about how air power can take its legitimate place in that struggle. The value of the heartland as a geopolitical concept has already shifted several times through history, as different modes of transportation have been developed, and it may do so again.

Alfred Mahan, for example, in reviewing the history of the eighteenth and nineteenth centuries, and the capacities of ships as opposed to land transport in that period, was convinced that the greater capacity and flexibility of movement by sea could prevent Russian expansion in Asia, and thus threw the weight of evidence against the heartland. H. J. Mackinder, writing at approximately the same time, but viewing geopolitics as a landsman, and noting how land transportation was improving at a faster rate than sea transport, reversed Mahan's interpretation. He gave to the heartland the great advantage of interior position, because it was the center of interior lines radiating throughout the Eurasian land mass, whereas the nations on the crescent were at a disadvantage by reason of their exterior position. Russia, in this view, would be able to shuttle supplies and reinforcements at will to any point that was attacked.

Both men recognized that the key to the situation lay in the fact that an interior position, geographically speaking, is of value only when the nation's transportation system and the extent of territory involved are such as to permit it to concentrate its forces faster than the enemy can do so, and thus to overwhelm him. Mahan believed that the sea lanes around Eurasia were really the interior lines strategically, because, as scattered as the maritime nations might be, they could nevertheless concentrate more rapidly than their continental enemies. The practical strategic value of Russia's interior lines, in other words, was far less than might be expected because in terms of space, capacity, time, and feasibility, sea transport outweighed land.

This controversy had barely opened, however, when the birth of the airplane threatened to overturn both views. Air power spokesmen were quick to point out that the traditional concept of heartlands would cease to be of value as soon as the long-range bomber was able to penetrate to all vital centers of the world. Indeed, it was predicted that as far as Russia was concerned, her vast spaces would be a source of weakness, because they could not be defended once air power could concentrate its blows from any and all directions.

Two things have occurred in recent history to complicate the neat and simple theories previously expounded. In the first place, Germany and Japan, which had both accepted unquestioningly the advantages of the interior geographic position, made swift and remarkable progress in extending their power outward from their own centers, only to be overwhelmed, in turn, by opponents who used the same roadways, oceans, and airplanes to strike back at them. It suddenly became evident that transport systems generally work both ways, and that the contest between the continental and maritime nations would be decided in large measure by how well each developed and employed its national power, and particularly what weight each gave its air, land, and sea transport facilities.

In the second place, even before the cold war began, and the bitter struggle between Communism and freedom had become apparent, the concept of the strategic value of the rimlands was suddenly projected onto the world scene by Nicholas J. Spykman. "Who controls the rimland," he wrote, "rules Eurasia; who rules Eurasia controls the destinies of the world." A concept which had been implicit in Mahan's writings at the opening of the twentieth century was thus made explicit. Indeed, the United States was soon to formulate its foreign policy approximately, if unwittingly, in Spykman's terms. Almost overnight, it became of vital importance to determine just what kinds of national power were going to be necessary, in order to prevent Red control of the Eurasian periphery. As the Communists extended their domination outward to China, to parts of Southeast Asia, to Central Europe, and toward the Mediterranean, and as the rimlands remaining in the hands of the free world shrank to an alarmingly thin line, it seemed to be a matter of touch and go whether shipping was still superior to trucking and railroads as determinants of the struggle, and the role which air power might play became increasingly important.

Here is where the view that we adopt of air power becomes crucial, for it will not only determine our manpower and resource policies and our basic strategy, but may well decide whether containment will succeed or fail? If we accept the air supremacy theory and place our faith in strategic air power, we are, for all practical purposes, giving up the rimlands for lost. It is assumed that the battle, if it comes, will take place over the polar regions of the north with the two major belligerents' cities, airfields, and launching sites as the main targets of attack. We are thus forced to place ever-heavier emphasis on the long-range bomber and the ICBM, to the detriment of land and sea power and to the exclusion of those components of air power essential to the successful execution of ground operations. Our position in the rimlands, already precarious even with fast, hard-hitting professional ground forces and versatile naval power, becomes untenable, and containment thereafter is dependent either upon Communist fear of our great deterrent, or upon our willingness to employ it whenever the enemy starts nibbling at the rimland. Not only does it become impossible to hold this modern version of no man's land, but, in the view of the air supremacists, it becomes unnecessary to retain control of anything outside of North America.

The contrary view, which holds that air power should be subservient to surface forces, of course considers the rimlands vital to the free world's position, but it perhaps gives insufficient weight to the fact that strategic air power in the enemy's hands makes it possible for him to jump the rimlands and strike directly at our vital centers across the seas. Even the composite view of air power may not be entirely adequate for the future, because it emphasizes too heavily the military or, more especially, the combat features of the airplane. It still tends to confuse the airplane with air power.

If we are to hold the rimlands, we are certainly going to need all the help from air power that we can get. It is more than likely, in implementing this containment policy, that there will be occasions when we will have to throw in troops, weapons, equipment, and supplies with great speed and in large quantities. What we will need for this purpose is a transport airplane comparable in emergency operations to the ship in peacetime commerce. This will permit us to concentrate instantaneously, and to sustain ourselves wherever we choose to attack or wherever the enemy selects to make us defend.

Just what effect nuclear-powered planes will have upon this struggle to control the rimlands, it is too early to say, but this much is evident: the Communist position on the mainland of Eurasia is a strong one, and the military air power which we count on to help us retain the rimlands may be equally useful to the Communists if they choose to assault the rimlands. The airplane as a weapon, in other words, can work both ways, as airmen well know. The speed of the airplane is, however, so great that it may enable us to neutralize the advantages that the enemy's interior position gives him in concentrating his forces. It is quite possible, in other words, that the nuclear-powered plane, when used in conjunction with allied sea power, may tip the scales in our favor.

It is important to note, moreover, that the maritime nations can make use of transport seaplanes to bring supplies across the oceans and land them immediately at points offshore, while the enemy will have the laborious and time-consuming task of constructing airfields before he can sustain large-scale operations. The significance of this for speed of concentration is enormous. It is one of the best examples of air-sea power, or the value of sea-based air power. By means of front-line airlift, intra theater air transport, and transoceanic air and sea logistics, the tempo of operations can be speeded up, and in the vast spaces of Eurasia this may well spell the difference between victory and defeat. Concentration in time and space, which is the essence of strategy, is the dominant feature of air power; it is a characteristic which we must be prepared to exploit not only in air battle, but in the whole range of military operations which demand mobility.

The best deterrent to limited war, of course, is forces on the spot, but since this is not always feasible for a variety of reasons - political, financial, and strategic our ability to deploy forces at any point on the rimlands, instantly, will serve almost as well. Certainly, if we can confront the enemy in a matter of hours instead of months at whatever points he decides to attack, we can stop him from presenting us with a fait accompli. Unable to obtain his objectives by military means, he will be compelled to turn to the field of political and economic combat.

Whether air power can ever contribute to this type of contest to the extent that sea power can, depends upon technological developments which are beyond the power of prediction. It is evident that if the rimlands are to be held, every aspect of air power strategic, tactical, and logistic must be given due weight. American national power, however, will not be fully developed, nor will the policy of containment be adequately implemented, until we view air power in its broadest aspects; that is, until it is based on the transport plane as much as on combat aircraft.

13

Nuclear Arms Control Agenda for India

Introduction

India's stated and implied goals for overt nuclearization are to counter the threat posed by Pakistan and China and to enhance the country's status in the international system. Pakistan's ongoing nuclear and missile programs were viewed as destabilizing because they changed the military balance in South Asia. The fact that Pakistan had been actively supporting the insurgency in Kashmir added to Indian concerns that a nuclear-capable Pakistan might try to aggressively resolve the Kashmir dispute.

In the case of China, India has a long-standing border dispute with that country, and China still occupies 38,000 square kilometers of Indian territory. Continuing evidence of a China-Pakistan collaboration on nuclear and missile technology issues made India believe that the country faced a deteriorating threat environment. There was also growing frustration with China's violation of Nuclear Nonproliferation Treaty (NPT) agreements not to transfer technology and with the fact that thirty-six years after the border war with China, not only was the territorial dispute unresolved, but China had worsened India's security environment by placing nuclear weapons in Tibet.

Politically, Indians see the development of a strong U. S.-China relationship as emerging at India's expense and condemning New

Delhi to second-class status in Asia. Militarily, the modernization of China's nuclear deterrent, the ongoing upgrade of its conventional armed forces, and Beijing's attempts to secure port rights in Myanmar are all viewed with concern by Indian strategic analysts. More recently, the revelation that China has acquired the designs for all the U. S. in-service nuclear warheads, as well as for neutron bombs, missile guidance technology, and satellites, made Indian defense planners feel that the country operates at both a qualitative and quantitative disadvantage vis-à-vis China.

Some analysts have argued that the development of an Indian nuclear capability could be an opportunity to create an "anti-China axis between the U. S. and India.... Behind the BJP 's bogus anti-imperialism and the American sanctions lies the prospect of a far-reaching alliance in a new Cold War. " Additionally, there was a perceived political advantage to be exploited. Indian strategic analysts have argued for some time that nuclear weapons and the means to deliver them are the "currencies of international power. " After the first test of the Agni intermediate-range ballistic missile (IRBM), India's leading security analyst, K. Subrahmanyam, argued that the possession of an IRBM capability meant that India's voice could not be ignored in any future negotiations on arms control and disarmament. After the tests, the Indian journalist N. Ram quoted an influential Indian expert as saying: "Tell us what we are and we will tell you whether we can sign. Guarantee to us that technology controls, which you apply as though we were a non-nuclear weapons state, will be removed. " Ram argues that the Indian government was claiming that if it was accepted as the sixth weapons state, only then would it be willing to abide by the provisions of the NPT. The Indian government hedged its bets, however, by suggesting that membership in the nuclear club did not automatically mean that India would become a hegemonic great power. Instead, its new nuclear status was to be used to "push for India's trademark vision of a world sans nuclear weaponry. "

In August 1999 India announced a draft nuclear doctrine that officially described the purpose of the proposed nuclear force and the shape it might take. The doctrine declares: "India's strategic interests require effective, credible nuclear deterrence and adequate retaliatory capability should deterrence fail. " The report states that

India will not engage in the first use of nuclear weapons. Nor will it use or threaten to use nuclear weapons against non nuclear states. In order to pursue a credible minimum deterrent, the country will develop forces that are "sufficient, survivable, and operationally prepared." It will also develop a robust command and control system under civilian control. The force itself will be "based on a triad of aircraft, mobile land-based missile and sea-based assets." Survivability will be enhanced by a combination of "multiple redundant systems, mobility, dispersion and deception." Space-based and other assets will be created to "provide early warning, communications, damage/detonation assessment." Finally, the Indian government will take steps to ensure the security and safety of nuclear weapons and will ensure that "unauthorized or inadvertent activation/use of nuclear weapons does not take place and risks of accidents are avoided." The draft doctrine also repeats India's desire for universal disarmament and global arms control and states that the country will continue working to achieve these goals.

In order to create such a triad, Indian analysts have drawn up plans for nuclear forces that range from 100 to more than 400 nuclear warheads. The late General K. Sundarji, India's chief of army staff in the 1980s, called for a mix of intercontinental ballistic missiles (ICBMs), IRBMs, and nuclear powered ballistic missile submarines (SSBNs) to provide India with a second-strike capability against both Pakistan and China. A later study by Brigadier General Vijai Nair called for a force of 132 nuclear weapons against China and Pakistan. One member of parliament, Subramaniam Swamy, has called for a nuclear force consisting of 400 IRBMs as well as airborne nuclear weapons. Bharat Karnad, who wants India to have 300–400 thermonuclear weapons with an ICBM capability, has argued for the most ambitious force structure. India has already taken the decision to induct the Agni 2 intermediate-range missile into service, and reports suggest that the country may test an ICBM named Surya in 2001.

It has been suggested that India also develop a naval capability that would serve as a second-strike force against both Pakistan and China. The need for a second-strike capability against Pakistan arises from the low warning time of seven to ten minutes that India has to react to an attack from that country. The need for a naval deterrent against China comes from the fact that China's major cities would be out of range of fighter aircraft and most Indian land-based missiles. The type of naval capability desired has varied. The proposed forces include using a sea-based version of the Prithvi (renamed Dhanush) short-range ballistic missile aboard Indian surface vessels; the building of a nuclear powered submarine the advanced technology vehicle (ATV) and the induction of an SSBN force; and the deployment of an indigenously produced sub marine launched cruise missile named Sagarika. The air-based leg of the triad would consist of nuclear-capable Jaguar, Mirage 2000, and Su-30 fighter aircraft. Recent reports also indicate that India is purchasing the Tu-22M Backfire intermediate-range bomber from Russia. Armed with a long range cruise missile, the Backfire would provide India with a second strike capability against the cities along China's eastern seaboard.

An ICBM-based force would, in addition, move India into the same league as the big five nuclear states by giving it the capacity to strike targets in the developed world. This would not only provide it with a true deterrent capability but would also, in theory, allow India to sit at the big table of nuclear "haves" when it came to discussing and formulating new arms control regimes.

Limiting Factors

Despite these ambitious goals, a set of constraints is most likely to determine India's eventual force structure. India's only real extraregional threat is China, and its primary focus of attention remains Pakistan. Further, to counter the Chinese threat would require a significant qualitative and quantitative increase in technology.

Developing extra regional forces would not only raise concerns among states in the area but also among the major powers. An Indian nuclear submarine, for example, would cause concern in both Southeast Asia and Australia. It would also alarm the major powers, especially China, which would, quite correctly, view the mission of the submarine to be one of targeting Chinese cities. Initially, therefore, India's nuclear forces are likely to be focused within the region.

Because these forces are primarily intra regional, the creation of large and complex force structures would be overkill. For deterrence to be achieved in South Asia, both Pakistan and India need to have the ability to cause unacceptable damage to each other, which can be achieved with far fewer resources than those used by the United States and the Soviet Union in the context of the Cold War. For Pakistan to cause unacceptable damage, it must have the ability to destroy four major Indian cities: Bombay, India's financial center; New Delhi, its administrative center; and Bangalore and Hyderabad, where India's high-technology industries are located. After the April 1999 missile tests, Pakistan's chief of army staff, General Pervez Musharraf, said that Pakistan would not try to match India in the number of missiles produced but would retain just enough missile capacity to reach "anywhere in India and destroy a few cities, if required."

In the Indian case, similarly, inflicting unacceptable damage on Pakistan requires the ability to destroy its two major cities Karachi and Lahore and its capital, Islamabad. The use of nuclear weapons in a war fighting mode along the Indo-Pakistani border does not make sense for either side because both have large population centers in that area. Further, it is difficult to draw the distinction between civilian and military targets, given their close proximity.

Economic constraints will also determine the size of the future nuclear force. After a period of high weapons imports in the early 1980s, there was a reduction in Indian defense expenditures in the late 1980s and through the mid-1990s, from 3.3 percent of gross domestic product (GDP) in 1987 to 2.8 percent in 1995. In 1992–1993, expenditures were as low as 2.4 percent. In 1996–1997, Indian defense expenditures were 3.3 percent of GDP. Defense expenditures will increase substantially once nuclear weapons, delivery systems, and command, control, communications, and intelligence (C3I) systems are included in the force structure. In fact, India did increase the defense budget in 1999.

If one were to use Subramaniam Swamy's calculations for a 400strong IRBM force, the cost of each nuclear warhead would be approximately $375,000 and the price of each missile approximately $3.75 million. When the cost of a C ^{3}I system was added, the total cost of the force was estimated at about $2.5 billion. Since Indian pricing on defense production tends to be deflated and hides expenditures under civil and military headings—one must assume that the cost of each warhead and missile may be significantly higher. Other sources, after taking a more conservative approach and incorporating the need for imports, put the figure as high as $40 billion. In terms of military expenditure, therefore, such a force would put severe constraints on the domestic economy.

Adding to the economic burden is India's continued need for economic assistance and foreign investment. The Indian economy is largely self-sufficient and will not feel the same type of economic constraints imposed by sanctions that the Pakistani one does. But India does require increasing foreign investment to carry out much-needed market reforms. It also needs access to Western technology to continue to modernize its economy. After the nuclear tests, investor confidence in the Indian market fell as the flow of investments almost dried up and the stock market plummeted by nearly one-fifth of its value (although one must add that the country was already in the middle of a recession). Maintaining access to high technology is crucial to India's developmental plans and to the modernization of critical areas of the economy. In fact, India has made the lifting of technology restrictions, particularly in the area of civilian nuclear technology, a precondition for signing the Comprehensive Test Ban Treaty (CTBT). Given that there is a need for developmental aid, technological transfers, and foreign investment, India will have to work with Pakistan to stabilize the nuclear environment in South Asia.

Finally, both India and Pakistan recognize the dangers of a nuclear confrontation and in the past have backed away from potential wars that could have escalated into full-blown nuclear conflicts. In 1986–1987 and in 1990, the two countries worked to defuse confrontations that might have escalated into broader military conflicts. In the May–June 1999 Kargil conflict, both India and Pakistan took pains to state that they were exercising restraint and seeking a diplomatic solution to the crisis. In light of these constraints, what type of Indian nuclear force is likely to emerge?

Possible Force Structure

It will initially be localized rather than extra regional; it will be smaller in size than the numbers that are being thrown around would indicate; and it is likely to be "recessed" rather than deployed (even though the Indian government has suggested that it will deploy). The draft doctrine makes a distinction between peacetime and wartime deployment. What that means in operational terms has yet to be determined.

Given how few targets India and Pakistan need to destroy, large forces will not emerge on either side. What is more likely is that both sides develop small nuclear forces having between twenty-five and seventy-five weapons to meet the threat posed by each other and the perceived Chinese threat to India. Both countries would make these warheads deliverable by aircraft and ballistic missiles. In the Indian case, there will probably be a sea-based deterrent as well. The naval deterrent is not going to be in the form of an SSBN force because it would be prohibitively expensive for India to build such a fleet. It would evoke concerns both among extra regional states and also among the major powers. Further, the development of the ATV has been plagued by delays, and the vessel is unlikely to enter service until the 2020s. Instead, India will probably develop the capability to launch the Sagarika cruise missile from its conventional submarines, giving it a limited naval deterrent capability. It will also gain a naval deterrent through the use of the Dhanush, which will be launched from surface vessels.

The first Chinese threat to India came from Beijing's conventional advantage along the Sino-Indian border and the fact that Chinese forces until recently had the advantage of a nuclear backup, whereas the Indian forces did not. The second threat from China came from the ability of Beijing to target Indian urban populations, whereas India, with its aircraft and intermediate-range missiles, could not attack Chinese civilian centers on the eastern seaboard. To match these threats, the Indian forces could use airborne weapons as well as the short-range Prithvi ballistic missile to stop a Chinese breakout through the Himalayan passes. Since India's policy is one of deterrence, blocking the passes would achieve the Indian goal of halting a Chinese attack and inflicting significant losses at the tactical level. The danger of China broadening the conflict could be countered by building a limited number of Agni 2 missiles, which are expected to have a range of approximately 3,000 kilometers and thus could reach cities in China. Targeting cities along China's eastern seaboard would require the development of a longer-range Agni or the use of the Surya ICBM.

There are serious limitations, however, to trying to implement such a strategy. In its first test, the Agni 2's range was revealed to be approximately 2,000 kilometers, not enough to target cities on China's eastern seaboard. Although India's Defence Research and Development Organization could develop the longer-range Surya, the time taken to develop the first two versions of the Agni and the international pressure not to test them give some idea of just how difficult it will be for India to develop a credible deterrent against China in the near future.

There has also been some discussion of using the polar satellite launch vehicle (which has a range of 8,000 kilometers) as a long-range missile. Doing so is technically possible, but the survivability of the missile in a conflict situation is questionable. As Dinshaw Mistry has pointed out, making a 170-ton missile road-mobile would be a difficult and cumbersome process. The alternative would be to build silos for it, thereby making it both easily detectable and vulnerable to a first strike.

Additionally, the nuclear force is likely to be what Indian defense minister George Fernandes called "recessed" forces or what the Indian government has alluded to as peacetime deployment. This terms mean that India would develop and flight-test its ballistic missiles and produce nuclear weapons but would not mate the warheads to the delivery systems. The warheads and the delivery systems would be stored separately and only assembled in the event of a crisis. Current deployment of the Prithvi short-range ballistic missile (SRBM) with conventional warheads along the border with Pakistan would suggest that such a policy is in place.

There are four good reasons to follow such a policy. First, past wars between India and Pakistan have resulted from long, drawn-out crises that gave both sides sufficient lead times to mobilize their forces. Second, both sides do not consider it likely that nuclear weapons would be used for a preemptive strike. In fact, both countries have recognized the irrationality of nuclear warfare and, as mentioned earlier, worked to defuse potential crises that would have escalated into a nuclear confrontation. Third, a recessed deterrence policy would provide assurances to the international community about both countries'; willingness to stabilize the nuclear situation in South Asia while allowing Pakistan and India to retain their nuclear arsenals and address what they see as genuine security threats.

Fourth, it will take India and Pakistan some time to build up their forces; 150 warheads will not emerge on either side in a short time. Instead, the process will probably take up to a decade, which means that both countries, for a fairly long time, will have small nuclear forces. The implications of this are best explained by Subrahmanyam, India's leading strategic analyst, who argues:

India and Pakistan have very small arsenals and for years to come they are not likely to cross the two digit figure. No one will use a tactical nuclear weapon since its use is an invitation to a much larger retaliatory strike. There are no proposals these days to have nuclear weapons on hair trigger alert. The countries are not likely to deploy their weapons lest they should lose them to even conventional strikes. Since the arsenals are small, extreme care will be taken not to lose those assets. India and Pakistan have the benefit of experience of the enormous blunders committed by the nuclear weapon powers and, therefore, have no illusions about nuclear war fighting or winning a nuclear war.

Keeping a small but recessed force, therefore, suits the diplomatic and military objectives of India as well as Pakistan.

In addition, the current policy of weaponization is being pursued by a particular regime, partly in response to the domestic political environment. In the foreseeable future, no Indian government could unilaterally abandon the nuclear and missile programs because of the fear of being labeled antinational and having surrendered to international pressure. But a succeeding government might decide to slow down the program by reducing budgetary allocations or weaponizing more gradually. Such a slowdown could be driven by economic considerations, international political pressure, and possibly the desire to develop better relations with Pakistan and China.

Such measures would limit force structures and prevent a costly and dangerous escalation of the nuclear arms race. Arms control measures at the regional level would provide some indication to the international community that both Pakistan and India were working toward managing their nuclear relationship and, therefore, becoming status quo nuclear states. A reduction in regional tensions would also help create a positive environment for international investment.

From a purely military standpoint, the willingness to implement arms control measures would allow for the continued development of strategic weapons systems. Without such measures, the pressure on India to cap, reduce, and roll back its nuclear program would be much stronger. However, entering into significant arms control agreements would give the Indian government breathing space to develop the weapons systems it requires for the country's security particularly long-range systems to counter the perceived threat from China. At the international level, having an active arms control agenda like the one discussed below would do more to enhance India's status than simply developing a limited nuclear force.

Regional Solutions

Two sets of problems must be dealt with following the nuclearization of South Asia: how to stabilize the post-test environment in the region and how to restrict the international fallout from these tests. After exploding their nuclear devices in May 1998, Pakistan and India sought to stabilize their nuclear rivalry. Initially, Indian and Pakistani proposals differed in their emphasis. India proposed a declaration of no first use of nuclear weapons by both states. Pakistan rejected the proposal because in a purely conventional conflict the asymmetry of forces favored India.

Pakistan made three different proposals: that Kashmir be made the core issue in resolving the broader Indo-Pakistani relationship; that India and Pakistan work toward conventional force reductions; and that both countries neither weaponize nor deploy their nuclear arsenals. As a consequence, negotiations between the two countries made little headway.

After almost a year of false starts, the two countries were able to initiate the first set of measures to formalize nuclear restraint in South Asia. U. S. pressure, domestic political and economic concerns, and the need to take steps to stabilize the nuclear rivalry brought the two countries to the bargaining table. At the February 1999 Lahore summit between the two countries, a memorandum of understanding was reached to develop a set of confidence-building measures. The two countries agreed to exchange strategic information about their nuclear arsenals and to give each other advance notice of ballistic missile tests. When India tested Agni 2 in April 1999 and Pakistan followed by testing the Shaheen and Ghauri missiles, both countries notified the other about the launch.

The two sides also agreed to engage in bilateral consultations "on security concepts and nuclear doctrines, with a view to developing measures for confidence-building in the nuclear and conventional fields aimed at the avoidance of conflict." In other words, both sides would exchange information on any C^3I measures they planned to implement to safeguard their nuclear arsenals. Further, both countries agreed to provide notice of "any accidental, unauthorized or unexplained incident that could create the risk of a [nuclear] fallout."

There were obvious gaps in the declaration. New Delhi's no first use proposal remained unendorsed, and Islamabad was unable to get conventional arms talks on the table or to get India to agree to a no-war pact. Moreover, Pakistan would have liked India to exercise strategic restraint (neither weaponize nor deploy its weapons), whereas India argued that it required a credible minimum deterrent against China.

In the regional context, therefore, the first steps have been taken toward formalizing nuclear restraint a sign not only of the awareness in both countries of the constraints they face but also of the success of lowkey U. S. diplomacy. But the one apparent weakness that emerges from this proposal is that despite an intention to exchange information on C^3I, little can be done until both sides have a secure capability to do this. Without a secure C3I system, long-term stabilization will not occur. Further, as the Kargil conflict has shown, significant progress in arms control cannot be achieved without a willingness to continue the political dialogue. In such a situation, keeping nuclear forces recessed or at least not on alert is the cheapest and easiest option for maintaining command and control over them.

Arms Control at the International Level

At the systemic level, the need for arms control comes out of the changed nature of the international system. India's nuclear policy originated in the Cold War, as did the assumptions about the advantages of going nuclear. The post–Cold War system, however, is a unipolar one, and nuclear weapons do little to overcome the overwhelming military advantage that the United States has as the remaining superpower especially in an era in which military operations have become high-tech and conventional weapons have a high degree of accuracy. Further, the international system is one in which nuclear weapons are neither the sole source of power nor the most effective means for achieving one's objectives.

Within the new international system, India will have to operate under certain given constraints. First, international pressure for nonproliferation will continue to mount, and with it the likelihood of sanctions will also increase. Second, a simple display of nuclear prowess does not translate into significant political gains like the acceptance of India as a great power by the international community. Indian hopes for using the nuclear capability to gain a permanent seat on the Security Council were dashed when the West made it clear that nuclearization would not be rewarded. Third, the West is unlikely to allow the economic collapse of either India or Pakistan because of the fear that such a catastrophe would force these states to transfer weapons technology for sorely needed hard currency. Fourth, a series of economic and military constraints make it likely that in an international system of first- and second-tier nuclear states, India will at best be a third-tier nuclear state.

Finally, India is an exceptional state. It finds it difficult to be an ally of the United States in the way that Israel or Britain is. There are too many differences between India and the United States about the future shape of the international system and how international politics must be pursued to allow such an alliance to emerge. Nor would any Indian government accept the status of a junior partner that such an arrangement would entail. At the same time, India is unwilling to be a rogue state like Iraq, Libya, or North Korea. The Indian government has consistently backed international law, supported the emergence and growth of international organizations, and sought to promote order and justice in the international system. Given these conditions, a carefully crafted arms control policy would reduce India's security dilemma, help improve relations with the remaining superpower, and allow India to play a significant role in world affairs something that low-level nuclearization does not allow.

Building a working relationship with the United States has been the first part of this process.

India-U. S. Nuclear Dialogue

At the international level the United States, as the nation carrying out bilateral negotiations with both countries, has called for the following measures. First, India and Pakistan must unconditionally sign the CTBT. Second, both must agree to join and further the fissile material cutoff treaty (FMCT) negotiations. Third, there must be a non deployment of nuclear weapons. Both countries also commit themselves to not transferring nuclear and delivery system technologies. Fourth, the two countries must enter into a dialogue to resolve the underlying causes of their rivalry. At the same time, the United States has moved away from its earlier position of attempting to roll back the nuclear status quo in South Asia toward, as one observer puts it, "a quiet, conditional acceptance of Asia's new nuclear realities."

In practice, the United States would need to partially accept the position of credible minimal deterrence that India has proposed. According to Indian prime minister Atal Behari Vajpayee, a credible minimum deterrent would include the following features: it would safeguard Indian security; it would be deployed; it would include a policy of no first use and nonuse against non nuclear states; and it would be deployed to ensure survivability. Further, the term minimal would not be quantified. Instead, its definition would remain flexible and be decided unilaterally by the government of the day.

The problem with this approach, however, is that because there are no finite numbers, it leaves enough uncertainty to promote an arms race in the region. To prevent a spiraling arms race, the United States has suggested that India spell out its minimum requirements in concrete terms, which would allow Pakistan to postulate its own requirements and stabilize the nuclear competition. Further, the United States has also pushed for both countries to demonstrate prudence and restraint in the development, flight-testing, and storage of ballistic missiles and nuclear-capable aircraft. Deputy Secretary of State Strobe Talbott has argued: "Having India and Pakistan stabilize their nuclear competition at the lowest possible level is both the starting point and the near-term objective of the U. S.

diplomatic effort. The Clinton Administration does not expect either country to alter or constrain its defense programs simply because we have asked them to."

The easiest way to do this, of course, is to go with a doctrine of recessed deterrence. It would remove some of the major concerns about accidental launches and unauthorized use of nuclear weapons and would also give both sides a chance to gradually set up a secure C^3I system. In fact, Indian officials have alluded to the low level of trust they hold for Pakistan, which makes it unlikely that the two governments could transfer sensitive data on command and control at this early juncture.

A more complicated issue arises from the political role that India sees nuclear weapons as playing. In the initial period after the tests, India had hoped to be recognized as a "declared nuclear weapons state" and to be invited to join an expanded United Nations Security Council as a permanent member. Since then, India has accepted it will not get this status, and instead, it has asked that it be called a de facto nuclear weapons state. But it is adamant about salvaging some political gains out of the nuclear tests. Getting the United States to lift the ban on civilian nuclear technology would be seen as a quid pro quo that would both assuage national honor and help quiet the domestic political opposition.

What makes it difficult for India to profit from its new nuclear status is the fundamental flaw in the belief that nuclear weapons are the international currency of power. Bharat Karnad, among other Indian analysts, has argued this point: "Nuclear weapons, first and foremost, promote strategic independence, are an attribute of Great Power and a manifestation of nuclear anti-hegemonism." For nuclear weapons to be attributes of great power, however, certain conditions need to be met. Like currency, one has to be able to display one's nuclear assets, which requires sufficient numbers to impress other great powers. Even if India were to make the economic sacrifice, face the wrath of the international community, and build 400 thermonuclear weapons, it would still be a second-tier nuclear state. If it built the more modest force that analysts like Subrahmanyam have called for, then it will be a third-tier nuclear power in a world where the first level of nuclear players are so far ahead that almost any Indian nuclear arsenal will seem minuscule. Only limited political leverage will be gained from saber rattling with a small nuclear force that does not have global reach.

A willingness to enter into substantive arms control discussions with Pakistan could, however, be used to try and gain economic and political concessions from the international community. One possible measure in this context would be for the international community to provide the right kinds of economic incentives to carry out negotiations. Economic incentives, rather than coercion, have worked to resolve regional rivalries and to provide disincentives to potential proliferators. The Egypt-Israel peace treaty continues to be bankrolled by the United States, and both Ukraine's possession of nuclear weapons as well as North Korea's budding nuclear program were terminated by providing economic incentives. Both the Indian and Pakistani economies require inputs of foreign investment if they are to carry out successful market reforms and modernize. The incentive of investments may well persuade both sides to carry out meaningful arms control negotiations and produce visible results.

One more immediate proposal would be to provide both countries with the means to safeguard their nuclear and missile assets from accidental launches and command and control problems. The Clinton administration has argued against transferring such technologies because it would only strengthen the nuclear capabilities of both states. But both countries have little experience with C^3I in the nuclear realm; transferring some technologies would help increase the safety of these arsenals. There has been a call from the region for the transfer of permissive action links to prevent the unauthorized use of such systems.

Future of International Regimes

The critical issue at the international level is that the proliferation of weapons in South Asia not have a spillover effect into other regions of the world, particularly the Middle East. Two questions are relevant here. Will India transfer nuclear and missile technology to other states? And does the present Clinton administration's approach of capping and reducing nuclear arsenals in South Asia (with the idea of a rollback being temporarily on the back burner) make for a viable nonproliferation policy?

To take the latter issue first, current U. S. policy aims to keep India out of the nuclear club of declared nuclear weapons states. Does such a policy make sense? Joining the club means abiding by the rules of nonproliferation, which include not transferring sensitive technologies, trying to prevent the spread of nuclear and missile capabilities, and working toward disarmament.

Until now, neither India nor Pakistan has transferred weapons technologies to other countries. India has had a nuclear capability since 1974 but has observed the non transfer clauses of the NPT, even though it is not a member of that regime. Further, it has been reluctant to transfer missile technology to other states. The Indian government turned down $150 million worth of missile-related business with Iraq and refused to enter into financial arrangements with Libya over the purchase of defense equipment. Indian officials argue that the country has abided by the provisions of the NPT far better than a declared nuclear weapons state like China, which has passed on both nuclear and missile technology to other countries.

Given that India has abided by the rules, entry into the club makes good sense. The alternative is that countries like India, Pakistan, and Israel form their own club where the requirements for membership may be less rigorous, leading to wide-scale proliferation. What is required, therefore, is an international legal regime or entity that admits the three countries with nuclear weapons Israel, India, and Pakistan as de facto nuclear weapons states. Doing so would make international regimes match the reality of the international system. This is a diplomatic initiative that the Indian government could pursue with the United States.

Alternatively, U. S. policy could continue to perpetuate the idea that these are threshold states. Not only is this policy unacceptable policy for both South Asian states (and certainly Israel, if it was included in these discussions), but it will also lead to policy measures that may actually aggravate the situation. The continuation of sanctions, for example, is likely to seriously hurt Pakistan's economy, thereby possibly forcing it to reverse its stand on nonproliferation in order to gain economic benefits.

Reshaping international regimes or creating a class of nuclear states will not lead to widespread proliferation in the international system. States that have foreclosed their nuclear option Argentina,

Brazil, South Africa, and Ukraine will not reverse their position because of the South Asian tests. International pressure and economic constraints will make such a reversal unlikely. Nor are the overwhelming majority of signatories to the NPT going to change their position because they have neither the resources, nor the political desire, nor the technological capability to carry out nuclearization. Only a small group of nations ruled by authoritarian governments continue to have the ambition to develop weapons of mass destruction. Drawing a distinction between democratic states like Israel, India, and till recently Pakistan and authoritarian states may address this problem: authoritarian states are less likely to abide by the rules of interstate relations, are more likely to engage in internal repression, tend to lag behind in the quest for market reforms, and are less observant of international regimes and treaties. In fact being a democracy may be made a precondition for accepting Pakistan's nuclear status.

As a status quo state, India could immediately propose and sign a treaty promising not to transfer nuclear and missile technologies. It could also sign the FMCT because its nuclear aspirations do not, and given its economic capabilities cannot, lead to competition with the major powers. Further, as the Indian security expert Sisir Gupta pointed out in the 1960s, going nuclear would never be enough because all it would do would be to make India the seventh, tenth, or twentieth nuclear state; the country would still have the status of a minor power. What mattered was coming up with plans to restructure the international system.

If India did not use its nuclear weapons but instead used the influence gained by possessing them to push for disarmament, then its role would continue to be one that was both moral and moderating in the international system. This could be achieved, he argued, by "taking steps to advance the world to a higher level of international order and collective security." As India is finding out, being a minor nuclear power gives little status in a world of major nuclear powers and, especially, in a new international system in which nuclear weapons are no longer the sole currency of international power. Responsible behavior, therefore, is the key to entry into the club of major states. Developing a substantive arms control agenda that helps make international regimes more relevant and brings about systemic stability would be the first step in this process.

14

Post-Nuclear Global Security

Introduction

Today it brims with military, economic, and environmental interpenetrations that crisscross all territorial boundaries. Under these new conditions, attempts to secure liberty through political separation and to maintain peace through preparation for war are bound to fail. Today, the interpenetration of societies means that people in the United States cannot enjoy liberty if people elsewhere threaten an attack against which there is no defense. Although the blessings of liberty remain as precious as they were in the late 1700s, the most effectual means of preserving peace have changed fundamentally. No one's life, liberty, or pursuit of happiness can be realistically achieved through sovereign disengagement from other countries. Indeed, as the chapters in this volume vividly demonstrate, the peace and security of one society can be achieved only through the active enhancement of the peace and security of rival societies. To be secure, all States need reliable international agreements or laws that specify what others may or may not do with their military power, their hazardous wastes, and even their economic resources.

Moreover, today as in 1776, liberty and security can be encouraged by making government accountable to people. But unlike the Eighteenth and Nineteenth centuries, when the blessings of liberty and peace could flow from nations taking separate diplomatic paths, today each government must be held accountable to all people affected by its major decisions, whether they live within that government's borders or not. No society, through separation or adversarial relations, can escape duties to others or protect its rights against others. Integration, not separation, is the yet largely unspoken watchword of the hour, the underlying need for common understandings that is addressed by alternative security concepts, policies, and institutions.

Meaning of Alternative Security

The terms "common security" and "comprehensive security" encapsulate the two fundamental qualities that distinguish the new concept of security from the old. Security can be achieved today only when rival nations hold it in common and only when people take a comprehensive view of security threats that encompass demographic, economic, environmental, political, psychological, and religious as well as military problems that jeopardize their future. For security to be held in common means that the United States, among others, must give sincere attention to maintaining the security of its rivals as well as of its own public and allies. If in the future a powerful adversary like the Soviet Union or another nation feels threatened, it might well increase its ability to threaten the United States or indeed engage in combat that, even if not directed against the United States, would almost certainly inflict severe security costs on the United States from radioactive fallout, global economic dislocation, climate change, or irresistible migratory pressures from unwanted populations. In an environmentally fragile age where nuclear and non-nuclear weapons of mass destruction may be grasped at almost any moment by many different nations, the United States cannot maintain its own security while ignoring or even inadvertently increasing the insecurity of other societies.

Indeed, the old and still-prevailing goals of national security, if pursued with traditional policies and preparations for war, actually destroy the prospects for common security. Today's exponents of more sophisticated nuclear and conventional arms those who endorse a $300 billion military budget for the United States and sincerely see themselves as the best of US patriots unintentionally constitute a new kind of internal enemy, probably more dangerous to the blessings of liberty than was the British monarch of the late 1700s. They unknowingly threaten US security by standing in the way of common security policies that are the only hope for war prevention in the long run.

The comprehensive nature of security needs constitutes the second emphasis proposed here. It pushes the conceptual change even further by moving it from military security, which in the past has been attained through a haphazard balance of military power, to safety from all of life's major threats, which in the future can be addressed through a wide variety of new international regimes and institutions governing global environmental economic, military, and human rights issues. To borrow some computer jargon, if in the past the militarily competitive balance of power system has been the default setting for interstate behavior in a relatively anarchic world, in an age of interpenetrating forces and military threats that cannot be repelled we need to utilize a more sophisticated program that can ensure that conflicts will be conducted through agreed-upon nonlethal procedures, that resource-sharing will proceed in an orderly fashion so the system will not crash, and that common values will be implemented efficiently through cooperative arrangements.

As a more comprehensive concept of security begins explicitly to incorporate non-military threats, the concept of "enemy" also changes. Those people who by their consumption patterns and driving habits intensify atmospheric carbon dioxide and thin the ozone could well constitute a more serious security threat to the United States than does a peasant movement demanding land reform in Central America or the unlikely threat of a nuclear attack from the Soviet Union. Yet many of the perpetrators of these largely unacknowledged, non-military security threats are found not in some distant land or in the embrace of a foreign ideology but, instead, among the governing elites and consumer segments of the United States and other friendly industrialized societies. Many of the contemporary Western security threats originate from within the Western camp, not from without. Thus, security proposals that dwell primarily on inter-adversarial diplomacy the traditional focus - simply are not bold enough to address the full range of security problems.

Envisioning Alternative Security Goals

Whereas prevailing security policies emphasize national sovereignty and military competition in the balance of power, exponents of alternative security envisage a growing ability to form bridging identities and institutions among nations. To be sure, new strategies of security enhancement must operate in the existing diplomatic framework, but they also must guard against merely perpetuating it, as traditional national security managers and inertial social forces are bound to demand. A powerful and effective future security policy must aim deliberately at establishing a new structure of international relations and a code of conduct that reflects the mutuality and comprehensiveness of security.

This aim can be sustained through the rough and tumble of politics with an intellectual map or animating vision of what must happen to enable the species to survive with dignity. That vision can be expressed conveniently as the ability to transcend five time honored boundaries, both tangible and conceptual, that now exacerbate all contemporary dangers to human life by fracturing the human mind, spirit, and community, and that consequently impede movement toward common, comprehensive security.

The first of these boundaries is the species-dividing boundary between nation-states, a physical-psychological boundary that produces easy acceptance of the "us-them" approach to economic and political life and the Good Guys/Bad Guys syndrome that Ralph White so compellingly shows to be a recurring problem in international relations. If this boundary and its accompanying "territorial discrimination" are not transcended with a heightened sense of human solidarity, collective violence and globally undemocratic political decisions will accompany the human race for as long as we are able to put off extinction. Intra-species national identities can be often are carried to extremes when they are wedded to military power, so that people in one sub-species group begin to think of other people with a different group identity as less than human. This separatist identity, which Erik Erikson called "pseudo-speciation" three decades ago, obviously facilitates killing and obstructs alternative security policies.

The second boundary is between rich and poor, both within and between nations, perpetuating economic inequity and widespread, unnecessary suffering and conflict. As Lloyd Dumas has demonstrated, gross economic inequities create the conditions that often give rise to violence and squander opportunities to employ economic relationships to build a strong structure of peace. The right kind of economic relationship, not just the extent of economic connections, produces peace. This point is illustrated by differences between the separatist, war-inducing function of French-German economic relations after World War I and the integrative, pacifying function of economic relations between France and Germany since World War II. Indeed, a balanced economic relationship not only helps keep the peace, it also is a more efficient economic relationship in the long run, of enormous importance in the coming age of scarcity. Yet the failure to narrow the gap between rich and poor in the years since 1945 demonstrates how widely people have accepted this self-destructive boundary in practice if not in rhetoric.

Third, an intergenerational boundary, which seldom is bridged, enables people recklessly to consume the planet's resources and overload it with pollutants without concern for future generations (or indeed for anyone's long-term needs) and without respect for generations who have gone before us. Short-term considerations dominate political decisions, corporate balance sheets, military decisions about how to achieve security, and even moral calculations.

Fourth, boundaries between the human species and the rest of nature must be transcended to increase respect for an of creation and to stop despoiling the biosphere. The prospect of population growth, deforestation, desertification, ozone depletion, and climate change constitute grave security challenges that no technological fix can meet.

Finally, commonly existing psychic boundaries that separate and compartmentalize the conscious and unconscious minds, the material and spiritual worlds, the word (professed values) and the deed (actual behavior), thinking and feeling, and "masculine and "feminine" values-all these increase the likelihood of war and freeze progress in transcending the four previously mentioned boundaries. These walls of the heart and mind reinforce authoritarian personality traits and violent power relationships, produce exaggerated fears of adversaries, inhibit flexibility and tolerance, and discourage the growth of universal identities on which future security will depend.

People who do not see the gaps in their picture of reality clouded by these five boundaries more readily employ violence against other humans and disregard nature. As the preceding chapters demonstrate, tendencies toward exclusivistic identification with one's nation, economic class, generation, and a single species, plus psychic habits characterized by separation and compartmentalization, distort people's perceptions and consequently limit their vision and empathy (White); unnecessarily constrict their political and technological decisions (Davis, Lynch); stifle a more efficient, less exploitative, and potentially stabilizing global political economy (Dumas); stifle the growth of robust legal norms, procedures, and institutions (Weston); unravel the weaving of more representative institutions and a more governable social fabric worldwide (Russett); and even warp people's most basic moral calculations (Coffin). On the other hand, by intentionally striving to transcend these boundaries of thought and behavior, people can envisage more powerful and effective security policies for the future. Ibis envisaging of a more secure and compassionate world calls forth the deepest meaning in the foregoing chapters and underpins all serious discussions about alternative security.

Identifying Guidelines for Alternative Security Policies

Explicitly announcing central principles or generalized intentions helps give overall direction and constancy to any nation's security policy. Once principles are publicly embraced, officials are less tempted to dismiss them for short-term opportunistic reasons. During electoral campaigns, discussion of positive principles protects candidates and the public from unscrupulous appeals to the most chauvinistic, xenophobic tendencies of the electorate. Endorsement of a principled alternative security policy by the public and by legislators also can shape legislation and constrain executive officials from deviating easily or covertly from what a more principled policy would consider prudent in the long run.

A principled policy, especially if widely known and increasingly honored throughout the world, also can serve as a corrective to the problem of "global partisanship." In a common security policy, where the entire species is considered the relevant constituency for decision-making, any national perspective, however bipartisan from the standpoint of domestic politics, is partisan from a global perspective. If electorates and officials become more aware of this problem because the same principles for security policy become touchstones for many countries, disputes are less likely to lead to violence. The use of principles as the basis for negotiations also aids enormously in settling potentially violent disputes.

By articulating and honoring principled goals for an explicit code of conduct, one nation can encourage and firmly press other governments to follow suit. International judgments by many nations, for example, articulating the same principles of fair play can more effectively moderate nationally partisan policies emanating from, say, Moscow, Tokyo, or Tripoli than can Washington acting, as inescapably it must, as a mere party not a neutral judge in a dispute with any other country. Also, where a principled approach is rejected by one's government, individuals and non-governmental organizations may nonetheless still be able to guide their own behavior or at least their discussions with fellow citizens by a new international morality that rises above the more traditional policies their national governments try to implement. Such a moral code reduces stress and expands strength in a social fabric that can increase the governability of the world.

Five basic principles or goals, informed by the five boundaries that impede movement toward common, comprehensive security, underpin a new international morality: demilitarization, reciprocity, equity, environmental sustainability, and accountability. Each is examined here in fight of the preceding chapters; but only the first, because of its special relevance, is analyzed in detail.

Demilitarizing World Society

The purpose of global demilitarization is to reduce the role of military power in international relations generally, in contrast to arms control which aims to limit weapons of one sort or another. Demilitarization may well include gradual reduction of arms, but it goes much deeper than arms control which, as it has been practiced since the 1950s, has meant merely the management of the arms buildup. Washington officials have departed little from the arms control approach despite the promise of the INF treaty and the enormous changes in the Soviet Union and Eastern Europe that virtually have taken away "the enemy" for the United States. In contrast, a demilitarization policy would implement concrete proposals leading to the eventual reduction of armed forces everywhere to the lowest level possible while still maintaining domestic peace and monitoring borders for purposes of regulating trade and immigration. It would be accompanied by the growth of world monitoring and security institutions, such as outlined in the preceding chapters, particularly in the contribution by Burns Weston, to replace over time the need of governments to rely on their armed forces for security.

Given the comprehensive nature of the security concept proposed here, demilitarization understandably extends beyond reductions of armed forces to policies designed to demilitarize the national and global economy, public education (to reduce denial, projection, and Good Guys/Bad Guys thinking in its content), and the social, psychological and religious habits that support the aging war system of international relations. Separatist, national military cultures would be gradually displaced by a culture of legal obligation and peaceful dispute settlement among diverse nationalities, able to flourish safely because they are protected by international human rights guarantees and no longer wedded to military power.

Assessing the Utility of War

Although most of the authors of the preceding chapters are sympathetic to a broad-gauged demilitarization policy, they may at points acquiesce too easily, albeit reluctantly, in a continued reliance on national military power for security. In my judgment, their analyses could profitably go farther toward considering systemic alternatives to war itself. An unflinching look at reality does not warrant retention of the war system, even with many reforms, beyond a transition period measured in decades, not centuries. Preparation for the transition, already pressed upon us by technology, must occur now if change is to be informed by reason and if it is to occur by transcending rather than deepening the boundaries that now threaten our safety.

The extent to which one can justify continued reliance on military power for security should be determined, if we are to be rational by the utility of war. Yet a plausible case can be made (1) that war is not useful (2) that it is not necessary, and (3) that preparations for war, which continue in many countries, undermine common, comprehensive security far more than they enhance it. If a sufficient body of evidence sustains these hypotheses, then alternative security thinking should not be satisfied with merely moving toward a minimum nuclear deterrent, nor even with the more ambitious task of abolishing nuclear weapons, but instead with abolishing war itself. Let us examine briefly each of these three propositions.

1. Is War Useful?

Even the most heavily armed countries have found their military power to be largely inapplicable for enhancing security in today's world. In so-called small wars against determined nationalist movements, both the United States and Soviet Union have suffered severe reversals. Vietnam and Afghanistan symbolize the ineffectiveness of vast military strength in a hostile environment against determined resistance. In both cases, the superpowers could hardly have achieved less through strictly non-military means than they achieved in their extremely costly applications of military might.

The prospects for benefits from larger wars are even bleaker; the likely destructiveness of major nuclear combat to one's own nation, one's allies, and the global environment cannot be justified by any conceivable political objectives. Both the US and Soviet presidents have said that nuclear war has no utility, can have no winner, and must never be fought. Influential religious organizations have called into question not only the use but also the possession of such weapons. Sensible responses to the threat of nuclear holocaust do not require fighting prowess; they require effective peacemaking to ensure that conflicts do not move to the brink of war.

A third category of violent combat, illustrated by regional conflicts such as the Iran-Iraq war and the East-West confrontation in Europe, includes wars that have become, quite simply, too destructive to constitute rational means for achieving desired ends. Even if a war in Europe could be limited to conventional arms, for example, these weapons are now so destructive that the conduct of the war would destroy the very populations whose protection is sought.

The declining utility of conventional war has become apparent to West Europeans, who, through the European Economic Community, have ended the age-old war problem between Germans and French and now are trying to fashion new and somewhat analogous approaches to enhancing East-West security. The declining utility of war is clear also to Soviet officials. And in the United States, time-honored realists who are independent thinkers have been so moved by an honest appraisal of the probable devastation of major war that they flatly say that not only nuclear combat but also conventional war among industrialized powers must be forever ruled out. Medium-sized wars also always run the risk of drawing in other countries, as occurred in the US flagging of Kuwaiti tankers in the Persian/Arabian Gulf, with the danger of escalating conflict. But most importantly, these conflicts are really not amenable to solutions brought by violence.

Are there exceptions to the rule that war has lost its utility? The US invasions of Grenada and Panama and the bombing of Libya may seem like exceptions to the rule, although on closer examination these actions may not in fact have served US security interests in the long run. In any case, alternative means were available for dealing with most if not all of the issues raised in each instance.

In the aerial attack on Libya, Washington associated antiterrorist efforts with US geopolitical and military interests and with anti-Libyan and anti-Arab sentiment. As a result, the policies drew criticism even from European allies and stood in the way of pining the widespread support that a less belligerent anti-terrorist policy could probably have achieved and that eventually will be required for anti-terrorist diplomacy to be effective.

In Grenada, US security was not in jeopardy. If any direct threats to the security of neighboring nations had arisen, they could have been met by multilateral peacekeeping operations. The invasion undermined an important interest the US has in maintaining a peaceful world order by upholding the principle that no government is allowed to intervene with military force to change another country's government. One such intervention can prepare the way for another.

Essentially the same points can be made about the US invasion of Panama to oust Manuel Antonio Noriega. However deplorable General Noriega's policies, there is little doubt that

Panama never constituted a threat to the United States. In the absence of a direct attack or a serious threat of direct attack on the United States or its significant interests, Washington could not legitimately claim to be acting in self-defense the only justification the UN Charter explicitly allows for the use of force. As in Libya and Grenada, the unilateral use of force probably undermined the US long-term interest in a peaceful, law-abiding world that is congenial to US security interests in the long run. In sum, collective violence has lost its utility in small medium sized, and large-scale war.

2. Is War Necessary?

Of course, critics argue that, although its utility may be low, war still may be necessary in some unusual circumstances. To determine whether this argument makes sense, consider the day-today security needs of the United States, recalling at the outset that it is not threatened by attack from anyone. Nonetheless, the United States and the world community have these security needs:

- to patrol and pacify tense borders in order to prevent brushfire incidents from erupting into war and to discourage the clandestine movement of arms and armed forces across international boundaries of countries where such problems have in fact arisen, e.g., Nicaragua, El Salvador, Honduras, Lebanon, Israel, Jordan, Iran, Iraq, Afghanistan, Angola, Namibia, Chad, Cambodia, and elsewhere;
- to discourage intervention by external powers in local and regional conflicts, such as the US and Soviet roles in Vietnam and Afghanistan, the Cuban presence in Angola, and the Vietnamese intervention in Cambodia;
- to prevent aggression by medium or small powers against one another, as Iraq against Iran, or Libya against Chad, or Middle Eastern countries against one another;
- to prevent ethnic or religious antagonisms from becoming violent, as between Greek and Turkish Cypriots;
- to implement peaceful settlements, such as the Israeli withdrawal from the Sinai and from part of Lebanon;
- to monitor arms agreements impartially and to publicize any violations in order to discourage them and to specify the nature

of the violations so as to avert overreaction by a rival as occurred in disputes over compliance with the SALT II treaty;
- to discourage the proliferation of weapons of mass destruction in the arsenals of countries already possessing them and of countries interested in obtaining them;
- to prevent space from becoming another source of military threat; and
- to provide impartial auspices to mediate or adjudicate disputes before they erupt into violence.

Not one of these needs can be met by more US military might or by unilateral US military pressure. If the Cubans have begun leaving Angola, it is because of multilateral political settlements on Namibian independence, UN monitoring of the agreement, and an end to other nations' military intervention in Angola. If clandestine gunrunning does not occur in Central America it will be because of impartial UN and the Organization of American States (OAS) observers at the borders and negotiated political settlements, not because of well-financed "Contras." If the Soviet military buildup is reversed, it will be because of non-military priorities in Moscow, detente, arms control, and international inspection, not because of US military threats. If war between countries like Iran and Iraq is prevented, it will be because the arms trade has not inflamed regional antagonisms and because the economic incentives for peace are greater than those for war; and if such wars do occur, external powers will need to avoid their escalation through conflict containment, dampening arms sales, and possibly the international protection of neutral shipping, not through introducing more armed forces in the region. If Pakistan, Brazil, and other countries do not manufacture nuclear weapons, it will be because the nuclear arms buildup of the superpowers has stopped and truly effective international rules against proliferation are enforced in every single country on earth.

By their nature, in sum, the day-to-day security problems most likely to prove nettlesome cannot be met adequately by national governments acting separately or unilaterally, no matter how strong they may be militarily. Modern security problems are made worse by unilateral uses of force. To manage them efficiently requires multilateral diplomacy and UN institutions that discourage external involvement by individual nations. Of course, the availability of multilateral peacekeeping and its imaginative extension in the future are additional reasons that war is not or should not be necessary. Peacekeeping under UN auspices, although certainly no panacea, can be more successful in serving most security needs than can uses of force by one nation or bloc of nations whose claims to legitimacy invariably are open to challenge.

Short-term security needs, then, can be well met by closing the door on national uses of military force and opening the door more widely for UN peacekeeping. True, there are some long-term security needs that UN peacekeeping, as presently practiced, cannot meet, the most obvious being the elimination of the danger of nuclear war. Others include the proliferation of non-nuclear weapons of mass destruction or the delivery of suitcase bombs. Yet national military forces cannot provide effective defenses against these threats either. Such security needs can be addressed in the long run only by policies that encourage global society to transcend the five boundaries noted above and to develop international norms for global demilitarization that are enforceable. This is what alternative security policies are designed to accomplish.

3. Planetary Militarization and Its Consequences

Preparations for war continue despite the declining utility of war and reduced military spending and deployments by the Soviet Union. In Washington, rather than developing a program for global demilitarization, for the prohibition of intervention in regional conflicts, and for the curtailment of covert operations, officials are implementing policies to develop more sophisticated nuclear weapons and new instruments for military intervention. These priorities are reflected in a report from a leading group of US national security managers, chaired by Fred C. Ikle, which concludes that the proper course for US policy is "discriminate deterrence," thus justifying the new weapons that the Pentagon wanted even before Gorbachev's more progressive initiatives (although now to be procured at a somewhat slower pace). The report encourages the United States to continue toward four goals: (1) developing new nuclear weapons and delivery systems, with greater speed and stealth, to give commanders more precision and ability to discriminate in the combat use of nuclear weapons; (2) increasing

US ability to intervene quickly and decisively in "low-intensity conflicts" in the Third World; (3) preparing an expanded role for covert operations; and (4) proceeding with the development of new weapons for warfare in space.

In accordance with this strategic direction, the Bush Administration is pushing forward with the development of space weapons; and, although Congress has authorized a smaller program than the Republican administration wants, Congress has refused to press for guarantees that space will be kept weapons free. Moreover, as a result of US insistence on keeping the nuclear door open, India, Israel Pakistan, and South Africa are rapidly enlarging their stockpiles of nuclear weapons material, with the first three trying to move from fission-based atomic weapons to far more powerful fusion-based devices. Argentina and Brazil can now cross the nuclear weapons threshold at will. Iran and Iraq, which are rearming to the teeth, already possess chemical weapons and missile delivery systems, and presently are eyeing nuclear weapons. Iraq, at least, seems to have taken the first steps in a program to build them.

In addition, the proliferation of non-nuclear weapons of mass destruction, especially chemical and biological weapons, proceeds at a quickened pace. The production of chemical weapons can occur throughout the world in laboratories originally constructed for the manufacture of herbicides and pesticides. Again, conventional military techniques are incapable of countering such threats. Only global constraints, multilaterally verified, can succeed in halting the further development of weapons of mass destruction.

The technology for producing aircraft and missiles has also been spreading ominously. The world's six most advanced newcomers to nuclear weapons technology Israel, India, Pakistan, South Africa, Brazil, and Argentina already possess advanced aircraft and are in the process of developing missile delivery capabilities if they do not already have them. China has sold a number of its missiles, with a range of 2,200 kilometers, to Saudi Arabia. And more than twenty Third World countries currently possess ballistic missiles or are making serious efforts to develop them.

The growth and further spread of military technologies profoundly endanger global security. Yet the prospects for stopping, for example, the further spread of advanced technology for rapid delivery of conventional chemical biological and nuclear warheads are exceedingly dim in the absence of serious steps to achieve a universal ban on ballistic missile testing. Yet such a ban cannot occur, nor can the spread of other advanced military technology be halted, as long as the major powers fail to demilitarize their own security policies.

Thus, preparations for war continue, undermining everyone's security in the long run. Given the imperfect functioning of the existing balance of military power and the proliferation of military technologies worldwide, we reasonably can expect that war some day will come again unless, that is, we transform the structure of international relations to provide sufficient governance at the global level to strengthen multilateral peacemaking and peacekeeping while negotiating and enforcing the demilitarization of national war making potentials.

Abolishing War

The failure to be sufficiently impressed with the inutility of war and of preparations for war has led many analysts to focus too narrowly on the dangers of nuclear deterrence. The focus is understandable because nuclear weapons are, of course, the most fearsome and least justifiable weapons in military arsenals. Yet, at its best, alternative security aims not at living without nuclear weapons; it aims at living without war.

Unlike many exponents of alternative security who seem willing to live with nuclear deterrence at minimal levels, I think it is unrealistic to believe that the threat of nuclear weapons can be reduced to a tolerable level by establishing a minimum nuclear deterrent, although admittedly it would be a useful step toward the abolition of nuclear arsenals. If the industrialized countries with nuclear weapons claim, as they do, a right to keep their weapons indefinitely, they cannot expect other less secure and more needy countries to give them up entirely and forever. This expectation, I speculate, simply will not be fulfilled. Either all countries must give up their nuclear weapons or all countries must accept that any country that wants them eventually will get them.

No amount of glossing over this stark reality should be allowed to divert those working seriously for an alternative security system. Advocates of demilitarization must examine carefully the widely held yet questionable belief that the world would be less safe with the abolition of nuclear weapons than with the maintenance of nuclear deterrence. This belief is based on one or both of two assumptions: that nuclear weapons have kept the peace for four decades and that someone could cheat on a total ban. Whereas hiding a few nuclear bombs is militarily insignificant in a world of thousands of weapons, hidden weapons could be exceedingly dangerous, the argument goes, where all countries except the cheater have dismantled their nuclear arms.

Although space does not allow a full development here, a plausible case can be made that these assumptions are erroneous and that efforts to abolish all nuclear weapons are desirable. The case is based on the following four central arguments.

First, as John Vasquez has demonstrated, there is no logical or empirical foundation for the belief that nuclear deterrence has prevented nuclear war. His analysis, as well as the recent work of other scholars, shows that the central element of US security policy over several decades may have had little to do with eliminating what was perceived to have been the main threat to US security.

Ralph White emphasizes that military deterrence, a much broader concept than nuclear deterrence, has prevented war in some cases. His is not, of course, an argument for the indefinite continuation of nuclear deterrence. Yet even the broader concept of military deterrence is not as compelling as often is presumed. Ralph White, for example, cites evidence that military deterrence prevented Hitler's Germany from making war between 1933 and 1938. Yet he may dismiss too easily evidence that military deterrence also brought a form of confrontational punitive diplomacy and unfriendly international economic policies toward Germany that created the political conditions that brought Hitler to power in the first place and that encouraged him and many Germans to make war in 1939. In other words, military deterrence did not succeed. In contrast, if the international community would have deemphasized military deterrence in favor of more conciliatory policies, such as proposed here, from 1919 on, it might have prevented the conditions that facilitated Hitler's rise to power or averted World War II altogether.

Ralph White believes that the case for deterrence against Hiders of the future is quite strong. Yet if a future Hitler possessed nuclear weapons, would the world be safe, regardless of how many other nations had them also? What would be an appropriate US response to a future Hider who might, during a showdown of threats, proclaim willingness to engage in combat to the brink of "nuclear winter," perhaps because of desperate conditions at home, unless certain "legitimate" wishes were met? If a future Hitler had nuclear arms, the threat of others to use their weapons against such a person might not be credible. Alternatively, their threat might be credible, but the future Hider might not believe it to be. Either way, nuclear war would result. If we generalize Hitler to be the stereotypic "crazy leader" or mad Caesar," nuclear deterrence offers no hope at all, because deterrence assumes a rational not a crazy opponent. If an opposing leader does not calculate rationally what other governments will do, or if that irrational leader does not agree with what others consider to be unreasonable levels of damage that ought to deter misbehavior, then deterrence fails.

Therefore, the international community must prepare defenses other than military deterrence against future Hitlers. It must ensure that future Hiders are unlikely to come to power, that they are unable to obtain nuclear weapons if they do come to power, and that they are unlikely to garner sufficient support from their own government and population to launch a suicidal war, because the firm yet non-threatening intentions of the rest of the world would be unequivocal. These goals can be achieved only by establishing a total ban on the possession of nuclear weapons in particular and by implementing alternative security policies in general.

Second, to the uncertain extent that military deterrence does reduce the likelihood of war, such deterrence can be carried out, at least against a government secretly trying to violate a nuclear weapons ban during a limited transition period, without resorting to nuclear weapons. Non-nuclear weapons are sufficiently destructive and precise to destroy any society on earth, and they can do so without nuclear fallout.

Third, cheating is not as large a problem as often claimed because intrusive inspection is becoming politically more acceptable and because intense developmental efforts can improve technological abilities enough to provide high confidence of monitoring an societies. In addition, modern communication, combined with individual citizens' willingness to report possible violations to a designated international monitoring authority, can produce high confidence of verifying a universal ban on nuclear weapons.

Fourth, the overall risks of nuclear proliferation, which attend any global posture less stringent than total abolition, exceed the low risk that any militarily useful secreting of nuclear weapons could occur without detection in a gradually demilitarizing world.

There is yet another reason for focusing more heavily on the war problem than the nuclear deterrence problem. Even if one seeks the total abolition of nuclear arms, it seems unrealistic to expect that the present members of the nuclear club will give up their nuclear weapons as long as conventional war against more populous adversaries remains a serious possibility. Even if additional countries do not rush to build nuclear bombs, the spread of chemical weapons, sometimes considered the "poor man's bomb," poses another frightening prospect that makes programs of partial arms control ineffective. Moreover, even if, for argument's sake, it were possible to dismantle all nuclear arms, nuclear knowledge still would pose deep-seated fears that some government might quickly rebuild such weapons during moments of tension assuming that is, that an effective global enforcement system had not already been established to prevent this possibility.

Thus, the nuclear sword cannot be lifted from our heads unless two conditions are met: (1) all countries must become permanent nuclear-weapon-free zones; and (2) all countries must drastically reduce their offensive, conventional military capabilities so as to preclude the possibility of large-scale offensive military operations. In our attempts to implement these conditions, the important focus must be not on any particular weapon or even claw of weapons, but instead on the gradual unrelenting, reliable reduction of the role of military force in international relations generally, until war itself has been abolished as an acceptable institution.

Restructuring Sovereignty

To focus on all forms of collective violence, rather than on nuclear deterrence alone, also highlights two other important points. First, the nature of the war-making function of sovereignty is changing. Indeed, it has changed so much that to retain any longer in national hands the ability to make war independently of any higher authority is to give up the overriding purpose of sovereignty itself, namely the ability to achieve the security of the nation. Second, the location of the residually legitimate function of the war making dimension of sovereignty police enforcement must move to the global level the only place where it can be reasonably carried out.

Apologists for traditional sovereignty in the balance of military power need not fear that they would be losing something that, at reasonable cost, they could retain anyway. Technological advances will continue to transform the nature of sovereignty, making separate, independent decisions less possible. Efforts to retain a war-making function of national sovereignty much longer actually will risk its extinction for many, if not all, nations through actual war or intimidation rather than bring its transformation into police enforcement for the good of the global community.

Moral Question

Above all, the moral issue must be faced more directly than any government and most people have done thus far. If nuclear weapons ever are used widely in combat they can only be characterized as genocidal and ecocidal weapons, indiscriminately inflicting death on many innocent people and the planet's life-support systems. In looking back now on the genocidal instruments of the Nazis, people almost universally conclude that individual Germans should simply have said "no" to paying for gas chambers, to constructing ovens, to operating these instruments of death, and to tolerating their very existence. Genocidal weapons, whether gas or nuclear, should be recognized as evil incarnate. One country's genocidal weapons do not justify another's. In any society where human rights are honored, no citizen should be required (or agree?) to pay for, build, transport, or tolerate the existence of instruments of genocide.

Non-Military Deterrence

While one side of a demilitarization program constrains the role of weapons, the other side increases the strength of nonmilitary influences. Non-military forms of dispute settlement, such as international regimes, legal tribunals, and boards of mediation, gain strength with frequent use. Diplomacy can be seen as an educational process in which every policy provides teaching and learning possibilities for the international community. The establishment of a permanent, individually recruited police force, for example, has an educational purpose that at this stage in history is as important as the physical security it can provide. Such a force could help people to understand that the global enforcement of community norms on behalf of the community is indeed possible. A permanent, individually recruited UN police force could establish a sense of global identity and impartiality that ad hoc forces made up of donated contingents from national armed forces cannot do as well. The agenda for building world security institutions includes also a global monitoring and research system, a non-interventionist security regime for smaller countries that could be protected by the UN, an environmental "security council" without veto.

The political flux in Europe presents historic opportunities to demilitarize and to create new common security structures in Europe. Yet, as of this writing, the Bush Administration plans to retain the East-West alliance structure and to bolster NATO by giving it some non-military activities. For the administration to ask that Germany remain within NATO, even if West and East Germany were to be unified, is a throwback to old, divisive habits of trying to obtain a military advantage through a peaceful gesture. Such policies are less useful than would be Washington's support for a plan to create an all-European security system in which the association between two parts of a permanently denuclearized and substantially demilitarized Germany would not threaten any country.

Demilitarization: In Whose Interests?

Exponents of alternative security need to pay close attention to the structure of interests and forms of learning that advance demilitarization. In this regard, they differ from advocates of traditional security policies who place almost all emphasis on military power. Also, they differ from the advocates of world federation in the 1950s who failed to develop an effective political strategy. A major problem in building consensus for change in North, South, East, and West is the failure to focus sufficiently on the need to develop concrete diplomatic and educational programs aimed not so much at policy reform as at the more fundamental task of transforming the international system itself. Current change efforts tend to be piecemeal haphazard, and half-hearted. If one or two countries negotiate arms reductions, or if several countries seek to curtail the arms trade, or if one region seeks to establish a nuclear free zone, but meanwhile the world continues to shy away from real reductions in the role of military power in the international system, even positive, incremental measures are bound, eventually, to fail.

Officials and publics need to recognize that arms are less the problem than the willingness of governments to use them and the failure of societies to build non-military means for security enhancement. Citizens and governments will not give political support to reducing their own arms very far if they harbor fears that other nations, even in the distant future, may threaten them. The US public's current receptiveness to arms reductions, for example, does not mean that it trusts the Soviet Union. On the contrary, at the same time that more than two-thirds of the voting public would like to move toward elimination of nuclear weapons, 68 percent continue to believe that "we cannot trust what Soviet leaders say, so we should proceed slowly and with caution." similarly, the public will not support substantial arms reductions if they believe that other nations, even though small and far away, may someday bring war closer. Although most US citizens, for example, no longer expect a direct nuclear attack on the United States by the USSR, 60 percent believe that smaller countries such as Pakistan, Israel, or South Africa "will eventually use nuclear weapons."

The solution to this problem is to develop and support a comprehensive plan that aims not merely at reducing arms but at reducing the role of military power in world affairs. To reduce arms and simultaneously take account of widespread fears of one's adversary, as is politically essential for every leadership to do, means that world security institutions, especially as outlined by Burns Weston, are necessary to help transcend the adversarial nature of today's security system and to provide additional reassurances that security can be enhanced by dependable global mechanisms. A successful plan must emphasize the growth of world security institutions as much as arms reductions so that alternative means for security enhancement are being erected at the same time that arms are reduced.

One of the most promising developments, of course, has been the series of remarkably progressive initiatives taken by Mikhail Gorbachev and the Soviet government. Almost without recognition by US officials, Moscow is asking the United States and the other great powers to strengthen international institutions and to revamp the customary code of international conduct so as to permit the mitigation of the pressing global problems that no national government can handle separately. Georgi Shakhnazarov, one of Gorbachev's closest special assistants, has given an extensive pragmatic rationale for Soviet determination to restrain traditional national interests that heretofore have disregarded the human interest. Writing in Pravda and elsewhere, he argues that Soviet policy must be guided by the need to raise the "governability of the world" to levels where global governance can manage global problems. Gorbachev's policies, he explains, ask "every member of the world community to place universal interests above class, nation, group, ideological or other interests." In a statement that is as remarkable as it is overlooked by the West, Shakhnazarov observes that even some of the main arguments against world government, which made sense for several decades after World War II, are "no longer there." In sum, Gorbachev's initiatives seek to bridle today's balance of military power with a growing web of international laws and organizations, until they function instead as a legally constituted balance of political power. In such a system, military power would be gradually relegated to a smaller and smaller role until war itself would be ruled out as a technique of conflict resolution.

Gorbachev's initiatives have put many internationalist items on the world's agenda that have not been there in years, if ever before. Creative space clearly exists in Moscow, not merely for reducing arms, but for the far more important goal of reducing the role of military power in international relations generally. Taken together, Soviet proposals and deeds offer more opportunity to institutionalize law and order in world affairs than has existed ever before in modern history, even more than what accompanied the close of World War II when Washington, Moscow, and the other Allied Powers created the United Nations. To this can be added the strong support for arms reductions, if not demilitarization, given by the nonaligned countries for many years. Repeatedly these countries have pressed for major steps in disarmament, including a comprehensive ban on nuclear testing. Moreover, Bruce Russett suggests that great promise may lie in the growing number of non-aligned societies in which democracy has taken root; fewer military governments means that more States will relate to each other without war.

Within the United States, which has tended to resist many arms control initiatives, especially since the beginning of the Reagan Administration, Thomas Lynch emphasizes the extent to which professional military people can be brought into the alternative security equation, by giving them important functions in defining new security structures. That is wise counsel as long as the goal remains clear: to dismantle not to remodel the existing war system in international relations. He also cautions that "no realistic alternative to nuclear deterrence can or should evolve without the active engagement of the world's militaries in its definition and execution." It is of course important to include the military in changes that affect them directly. Yet is it accurate to suggest that no realistic alternative to deterrence can evolve without the active involvement of the military? Are not alternatives evolving already, with every person, church, municipality, and country that rejects nuclear weapons helping to bring these alternatives closer to the foreground? If sufficiently widespread, might not these efforts convince others to accept a nuclear-free outcome, even though their fears prevent them at first from embracing it on their own? Surely supporters of alternative security need not wait for the architects of nuclear deterrence to dismantle it any more than slaves and abolitionists needed to wait for plantation owners to dismantle slavery. To be sure, people should be civil toward their political opponents, but they need not delay pressing for major change until the last holdout participates in planning the change. Do "sovereign military forces" have as much right to threaten societies with instruments of genocide as a sovereign people have the right to insist that genocidal weapons be abolished?

Thomas Lynch seems to argue that, even if the military arguments against alternative security have little logical basis to them, exponents of alternative security should scale down their proposals by not calling for actions that would drastically cut military budgets or personnel. He feels that such cuts will arouse the powerful military bureaucracy to oppose them. True, it may be easier to demilitarize the military than to dismantle it. But perhaps Thomas Lynch 's argument concedes too much. If military officials cannot justify on security grounds everything that they spend and deploy as, for example, they could not do in deploying multiple, independently-targetable re-entry vehicles (MIRVs) then no one serves the country's security by moderating policies just because the military bureaucracy might oppose them. As occurred in the MIRV case after Moscow deployed multiple systems also, the result would be lost security.

Indeed, Paul Kennedy has demonstrated the folly of such a strategy even on strictly economic grounds, showing that mature great powers and their military bureaucracies usually make the wrong decisions for their own good. They overextend themselves economically to procure military power until eventually they undermine their own strength. The policies of the White House and Congress over the past decade, running up the largest deficits in history primarily to buy unusable weapons, seem to confirm that, unless it alters its course, the United States will not save itself from this historical pattern of unnecessary demise.

The burden of proof, then, must be on those who propose more military equipment to demonstrate that their proposals do not undermine security. Militarization programs must be compared with demilitarization programs to determine their relative contributions to the full range of security problems, including the threat of economic losses, nuclear proliferation, global warming, and the prospect of "nuclear winter" if nuclear combat should occur. After lengthy study, John Tirman has reported that "breaking the grip of the military estate on America's policy-making weapons production, and value system would go further than any other measure to reorient national security to more appropriate ends."

This task need not wait for other governments or even for Washington to act. Individuals and non-governmental organizations can work at it through local schools, churches, and the media.

Indeed, alternative security thinking underscores that the control of the military by persons who do not operate from a military mentality is different from and more important than civilian control of the military. Years ago civilian control seemed a solution to averting the dangers of militarism. But today many of the civilians most commonly selected to lead the national security establishment have adopted military values. In fact, they often urge military interventions where professional soldiers dutifully agree to go but know in advance that a quagmire awaits them. Permanently high levels of military preparedness mean that militaristic minds, whether civilian or not, usually control the armed forces. They are advocates of more military power rather than of a decreased role for military power. Thomas Lynch and others give insufficient attention to this problem. As Tirman concludes, "if the grip of military values on society is not loosened, the possibilities for innovative leadership essential to common security are very dim."

Of course, policies that reduce the military budget and bureaucracy are bound to be politically troublesome, as Thomas Lynch correctly reminds us. This is why alternative security policies include legislation to manage economic conversion. Lloyd Dumas has shown how imaginative arrangements can adjust the military's economic and bureaucratic interests without leaving people unemployed even after military budgets are sharply reduced. Such proposals recognize the goodwill that exists within parts of military organizations and attempt to build political support within the military for what needs to be done. This seems more useful than moderating requests for demilitarization because vested economic and military interests may oppose them.

Military support may grow also from a recognition that there are many important future tasks that military personnel can fulfill better than others. Defense ministries could provide extensive monitoring and inspection personnel to improve the ability to verify norms restricting the manufacture and possession of arms. In addition, expanded research and development programs to construct better verification equipment are needed. Security enhancement activities would also include policing environmental standards to avert accidents at sea or in space and taking measures to clean up or otherwise recover in the wake of environmental or natural disasters.

Other tasks might include volunteering for a permanent, individually recruited UN police force once established, just as peacekeeping has become a natural and growing part of the role of the Nordic armed services, for example, where military professionals have come to accept UN peacekeeping as a significant part of military life. National military personnel would, of course, continue to police borders for the purpose of regulating commerce and immigration even in a demilitarizing world.

To military people genuinely committed to the enhancement of security, measures such as these would be attractive as long as it is clear that, by curtailing traditional military roles for one's own country, similar limitations would eliminate the threats that any adversary might pose to one's own society. Indeed, the military's highest duty is to do the utmost to reduce threats to the security of the homeland. Helping to implement a prudent program for global demilitarization could be such a high calling.

Honoring Reciprocal Rights and Duties

We turn now to a briefer discussion of the remaining four principles of alternative security policy. The principle of reciprocity, the first of these, is perhaps the most fundamental and already widely endorsed principle on which an alternative security policy can conveniently rest. Although reciprocity is frequently violated in practice, its universal endorsement by governments, regardless of ideology, nationality, or religion, provides a basis for attempting to hold governments accountable to a fundamental ordering principle. If rigorously applied, no government could with impunity insist on a right for itself that it would not willingly grant to others, or claim a duty for others that it would not accept for itself.

If seriously implemented, the principle of reciprocity alone could help to eliminate most wars. For example, if the United States denies that the Soviet Union or Iran, Libya, Syria, Vietnam, and other governments have a right to support armed insurrection or to finance the clandestine movement of military forces across borders, then, following the reciprocity principle, the United States could not claim that right for itself, as it has done in attempting to overthrow the Nicaraguan government. If the United States does not want to extend to other governments the right militarily to invade their neighbors to install new governments, then Washington must not claim that right for itself (even for the purpose of ridding Panama of Manuel Noriega). Alternatively, if the United States claims a right, as it has done under the Carter Doctrine, to use force unilaterally to maintain access to the oil fields of the Middle East, then may poor countries claim a right to use force to gain access to the corn fields of the Middle West or to encourage a migratory "invasion" of starving people to obtain food in a country that, by most measures of international justice, consumes a disproportionate share of the world's resources?

To install reciprocity as an operational policy principle can also drain heated ideological hostilities or religious fanaticism from many conflicts, because disputing officials can then focus on principles of good conduct rather than on the ideological goals of the other. Ideologically diverse governments can live peacefully with one another as long as they understand and explicitly design policies to respect this fundamental principle. Reciprocity and "realistic empathy" reinforce each other and provide a check on people's unconscious rejection of evidence that conflicts with "unconsciously cherished images."

In addition, reciprocity provides an essential antidote to one of the most serious dangers of a principled foreign policy: the tendency to assume a moralistic attitude, based on the assumption that one's own policies are more virtuous than the policies of others. As well schooled realists know, governments characterized by "legalismmoralism" often insist on their own way, propagandize for the "correct" point of view, and threaten other nations who obstruct their policies. Because evaluating respect for reciprocity can be controversial as illustrated in US and Soviet claims alleging violations by the other of the ABM treaty, compliance with reciprocity can be enhanced if its meaning is authoritatively interpreted by third parties whenever possible. Moreover, the implementation of complementary policy principles, such as democratization and demilitarization, discourages the possibility that a policy inspired by a moral vision will take on a self-righteous or aggressive tone. The recommendations in this volume by Ralph White for empathy in policy-making, by Burns Weston for wider application and impartial implementation of legal norms and processes, by Lloyd Dumas for economic fair play throughout the global arena, by Bruce

Russett for recognizing the peaceful potential in nurturing democracy globally, and by William Sloane Coffin for honoring the contribution of moral principle to prudential politics -all undergird the principle of reciprocity which in turn constitutes the bedrock of alternative security. Alternative security means nothing unless it includes an active concern for the security of one's adversary, because there is no path to one's own security other than through the door of reciprocal expectations that adversaries will reduce their threats in return for one's own threat reduction.

Achieving a Sustainable World Society

Without deeper respect for nature, a life of dignity for the human species cannot continue on planet Earth. Environmental issues pose planetary dangers of such a magnitude in scope and severity that they constitute the most serious long-range security problem in the world today. As the World Commission on Environment and Development (commonly called "the Brundtland Commission") concluded, life support systems for the entire human species face profound and uncertain threats from pollution, resource depletion, population pressure, and species extinction. Although these serious security problems can be addressed in part within local and national contexts, none can be treated effectively through the traditional security instruments of military strength or even through the traditional diplomatic instruments of bilateral diplomacy. People now need decision-making authority that is globally binding. Only the transfer of massive financial resources and brainpower from military to environmental purposes and only truly cooperative, binding, multilateral legislative efforts to protect the ozone layer and halt climatic change induced by global warming will enable the species to survive and enhance the dignity of life for future generations.

Although the problems seem at first to be overwhelming, the World watch Institute has estimated that $77 billion a year over a single decade could reverse adverse environmental trends in four key areas: protecting topsoil from further erosion, reforesting the earth, increasing energy efficiency, and developing renewable sources of energy. This cost amounts to only 8 percent of world military spending each year, with the benefit of laying a foundation for subsequent steps to end hunger and avert global warming. Alternative security policies could redirect large portions of global expenditures on research and development for new military technologies, estimated at $100 billion in 1986 alone, into developing new energy technologies, increasing agricultural productivity, enhancing pollution control and improving human health. All the world's governments combined presently devote fewer funds to these activities than what they spend on developing new military technologies.

Both the exponents and critics of alternative security policies can benefit from studying the exciting possibilities for managing environmental problems that have arisen from the willingness of some governments to reformulate the nature and relocate the exercise of sovereignty. Two dozen governments meeting at the Hague in March 1989 called for a strong international environmental institution within the UN system to render binding decisions even when unanimity cannot be achieved among all members. Disputes that could not be resolved through negotiation would be referred to the International Court of Justice. Cooperation in one security area, such as managing environmental problems, of course suggests models and establishes trust for carrying out alternative security policies in other areas, such as arms reduction and global police enforcement.

Achieving Equity

The guideline to advance equity throughout global society arises from a moral desire for more justice and a pragmatic need to achieve more economic rationality, international cooperation, and willingness to sacrifice for the good of all. A more equitable distribution of economic and political resources would contribute to development programs that meet the needs of all people, as well as reduce the political power of military establishments. In addition, it would help to deprive adversaries of the opportunity to maintain what Ralph White calls a "diabolical image" of their opponents.

In part because the world community has not eliminated glaring inequities, the world economy now functions so inefficiently that it damages the interests of all nations, rich and poor alike. The North's industrial capacity remains underutilized while the South urgently needs goods the North can produce. This inefficiency helps to perpetuate the inequities that cause the inefficiency in the first place, creating a form of global apartheid in which North America, Europe, Japan, and a few other countries live in relative affluence while one-half of the species lives in varying degrees of poverty. These inequities constitute a security threat for three reasons. First, poverty and inequity give rise to violence and militarism. Military governments and exponents of covert activities, whether governmental intelligence agencies or underworld drug dealers, often take advantage of poor societies. Richer, more powerful organizations manipulate the poor, inviting both the violence of authoritarian repression and revolution against it. Such conditions also encourage military interventions by others.

Second, poverty and inequity undermine the growth and stability of democracies. In societies facing desperate conditions, extremist political leadership or authoritarian military governments often come to rule. Poverty also stimulates unwanted population growth that would be reduced if more equitable, higher standards of living were attained throughout the world. Third, poverty is a cause as well as a consequence of environmental decay, which in turn threatens national and international security. Poor societies try to cut development costs by accepting low environmental standards for polluters. They may destroy rain forests to earn cash, even though the forests produce oxygen and absorb carbon dioxide, essential functions for people everywhere. Poor people, in desperation, also may destroy plants and trees while scavenging for food and firewood. Their herds may overgraze marginal pastures and add to the spiral of deforestation, desertification, and global warming. Overfishing, overgrazing, desertification, and loss of topsoil contribute eventually to a declining standard of living for the species everywhere. As a result, poverty and inequity should become security concerns for the rich even if they have little moral concern for justice or for the well-being of the poor. No national government, no matter how enlightened, can progress toward a less militarized, ecologically healthy, and politically cooperative world society unless today's gross economic inequities are overcome. Toward this end, governments should consider the following illustrative measures:

the development of economic conversion plans to reduce domestic opposition to demilitarization from corporations and workers benefiting in the short run from military contracts and to redirect resources to achieve other security goals such as abolishing worldwide poverty, sharing burdens fairly to achieve environmental protection, and providing sufficient prosperity to reduce population growth;

- serious international programs to conserve energy and other resources for the purpose of combating hunger, disease, underlying poverty, and environmental threats to security;
- automatic transfer payments from rich nations to poor nations, as recommended by the Brandt Commission, for example, which urged that all States contribute to a capital fund for development, based on a sliding scale related to national income and payable through what would amount to an element of universal taxation,"[57] in part because appropriate economic development can enhance national and international security; and
- more equitable representation of poor countries in the international monetary system and the development banks because such representation would help to establish more stable exchange rates, bring symmetry to the burden of balance-of-payments adjustments, expand international liquidity, deal with the debt that keeps many societies buried in poverty, and thereby contribute to the overall stability of world society.

In addition, a greater degree of equity within and among countries would increase everyone's stake in avoiding traumatic disruptions of world politics and encourage all States to support politically stable, representative international institutions, the fifth principle for framing alternative security policies.

Democratizing World Society

The democratization of world politics is no less important for alternative security advocates than the demilitarization of world politics. At the bare roots, conflicts grow into wars because some government mistreats or threatens to mistreat some people and other governments oppose the mistreatment, with arms when they deem it necessary. But if democratic processes can be nurtured at every level from the family to the planet, we are likely to experience far less mistreatment and to enjoy a far more peaceful and economically equitable world, as well as a world more respectful of human rights and nature. If human rights protections against mistreatment, whether it occurs as a result of domestic authoritarianism or international aggression, can be established with the help of more democratic international institutions, then no legitimate rationale for war would remain. Thus democratization and demilitarization reinforce each other.

Increasing Governmental Accountability

The most potent antidote to war is to increase governmental accountability to all people affected by government decisions and to construct international institutions to ensure that people with severe grievances have a court of appeal outside their own country. An effective alternative security program intentionally tailors all of its policies to implement the principle of accountability at all levels of social organization international, national and local.

Our focus here is primarily on the importance of increasing governmental accountability to people who, regardless of their nationality, are affected by the political economic, and environmental decisions that various governments make, regardless of the national capital in which a decision is made. In concrete terms, if the steel mills in Gary, Indiana, cause lake-killing acid rains in Ontario and Quebec, then the environmental decisions by Indianapolis and Washington to govern the mills and other smokestack industries of the Midwest must be accountable, in this regard, to the people of Canada. Yet the principle of accountability, so important to the first Continental Congress and to those who founded the United States, has been virtually ignored in recent years while a growing number of decisions that affect the lives of US citizens have surfaced in Japan, West Germany, the Soviet Union, and elsewhere.

Like it or not, to maintain a democratic way of life domestically, the democratic principle of accountability must be implemented internationally. As interdependence increases, the number of decisions that affect the citizens of one country, yet which occur outside that country, will increase also. For example, if the political, economic, environmental, and human rights decisions that affect, say, the US public increasingly occur outside the United States but representative global governance does not increase, then the degree of democracy for US citizens declines, even though domestic democratic forms, such as elections and a free press, continue to function. Of course, precisely the same problem exists for people in other countries, only for most of them the degree of democracy is smaller to begin with, either because their own domestic institutions are not democratic or because they lack the power and wealth that has enabled the United States to be well represented -arguably over-represented in world diplomatic councils.

Most citizens of existing democracies are not aware of the extent to which they are letting democratic fife ebb from their societies, because they assume that patriotism means concentrating sovereignty at the national level. Although they could, alternatively, support popular sovereignty in which people allocate political authority at local national and global levels according to its functional utility, many citizens are unnecessarily fearful even hostile, to internationalist solutions to contemporary problems. As George Bush declared when campaigning for the US presidency: "I will not turn one ounce of our sovereignty or of our leadership over to the United Nations. We must continue to lead the free world."

Such a proclamation reveals a false consciousness. To think that one protects democracy and freedom by concentrating decision-making power in separate national governments does not make sense in a world where political decisions and influences that affect one's life occur outside one's own national institutions. Analogously, a family may establish an internally democratic structure of relationships among its members, but if it cannot participate in the governance of its neighbors or if it happens to live in Nazi Germany, it does not live in democracy.

The physical security of everyone in the Northern Hemisphere depends, from time to time, on military decisions made in Moscow, Washington, or perhaps Tel Aviv or a Twenty-first Century Sarajevo. And in today's world the economic security of most societies depends on decisions made in Tokyo, Brussels, and Washington, as well as in their own national capitals. Alternative security underscores that the society relevant to anyone's life is now global as well as national and local. As a result, patriots of democracy deceive and potentially destroy themselves if they insist on being patriots also of a sovereignty fossilized exclusively in the territorial State. Alternative security seeks to remodel sovereignty rationally to overcome the inadequacies of existing world security arrangements.

Toward this end, most nations would benefit from establishing more democratic global institutions as soon as possible. The reliability of such institutions would then be firmly established by the time that governments almost certainly will need to depend on them. In delaying these developments because of reluctance to embrace multilateral diplomacy, the United States is pursuing a perilous course toward the year 2000, when the industrialized countries "will be home to only 20 percent of the world's people." Given these realities, as well as the moral imperative to enhance respect for all peoples' human rights, it is essential to establish strong protection for minorities wherever majorities rule, whether nationally or in evolving global institutions. Such protection will encourage minorities to accept majoritarian procedures globally and help to dampen hostilities among ethnically diverse populations nationally. Perhaps the only way that people in fragmented multinational societies will be able to handle their nationality problems fairly and non-violently is (1) to institute international guarantees for minorities and (2) to reduce, through global democratization and demilitarization, the differences that ethnic groups seeking more autonomy might experience between, on the one hand, having their own independent nation-state and, on the other, constituting a province of a national federation or confederation that operates responsibly in a law-oriented world community. The implementation of these two conditions would also enhance international peace.

Fortunately, instruments to protect important minority rights exist already in the UN human rights covenants and other human rights treaties. If ratified, they enable citizens to bring grievances against their own governments within designated international settings for possible redress of injuries. The United States should not delay the ratification of these treaties any longer. Once a party, Washington could more effectively support efforts by the UN Human Rights Commission, regional institutions, and non-governmental organizations like Amnesty International to monitor the performance of all governments. The advancement of human rights can of course stimulate additional political strength among persons who are committed to alternative security policies throughout the world but whose influence has not been previously felt.

Expediting Global Decisions

Stronger, representative global institutions are needed also because traditional diplomatic approaches take too long to produce important decisions. Prolonged negotiations, often hampered by the veto power of some States or the non-ratification of treaties by others, limit the world community's ability to anticipate crises before they become acute and to ensure a globally unified response in relation to them. Yet global environmental threats, to take only one example, increasingly need to be anticipated and managed before they occur. In the military sphere, anticipatory decisions to control space weapons, radiological weapons, and other weapons of mass destruction require building international consensus and truly regulating what all nations do militarily. Muddling through with traditional balance-of-power procedures is too crude and weak an approach to ensure reliable security policies.

Ending Covert interventions

Alternative security thinking also aims at eliminating covert activities. Stuffing ballot boxes, manipulating elections, bribing officials, assassinating unwanted political leaders, plotting coups d'état, and funding "secret" wars all such activities are unnecessary for security and unjustifiable in a principled security policy. They violate democracy and self-determination. They often contribute to violence. As for reciprocity, what country would allow these activities to be conducted in its own or its allies' territories? If some international interventionary action clearly is warranted to rectify a desperate situation, then it can be taken openly. If the need arises for international policing, this could be done through multilateral peacekeeping. Unlike covert interventionary operations, the covert gathering of intelligence might well continue. But with the creation of an effective global monitoring and research agency within the UN system, no government would need to rely as heavily on its own intelligence sources as it has in the past.

Nurturing "People Power"

The breathtaking successes of "people power," most recently in Eastern Europe, are another source of encouragement for building new coalitions of political support to implement alternative security policies. Recent progress in removing autocratic governments around the world can, if properly nurtured, enormously increase the prospects for institutionalizing more democracy at global levels of governance. World peace will benefit because authoritarian governments are less likely to serve the interests of their people, including their security in the broadest sense, than are more popular, responsive governments. In short, the spread and strengthening of "people power" is conducive to global security.

Democratic societies appear to nurture more inclusive forms of identity toward each other than they do toward authoritarian societies. At least between democratic societies, progress in democratization is a way of combating "pseudospeciation" and ending people's exaggerated fears of opponents that often lead to conflict and war. Democratic procedures facilitate the growth of realistic empathy. They enhance people's sense of inner strength as individuals and make spiritual renewal a genuine possibility. The contagion of anti-autocratic movements that swept Eastern Europe during 1989-90 illustrates the power contained in people who decide to exercise responsibility for the shape of their government.

The power of civilian resistance as a means of national defense, although it is different from mass action to oust an indigenous authoritarian government, pins some credibility from the way that repressive governments were removed from Eastern Europe in 1989. In assessing the utility of "people power," Thomas Lynch claims that it could not succeed against ruthless governments. Yet in both East Germany and Romania, where the respective heads of government ordered well-trained security forces to kill as many demonstrators as necessary to "restore order," the governments were later forced to abdicate. The order never was carried out in Leipzig after lower ranking officials, confronted at the scene with hundreds of thousands of protestors, reluctantly reversed a written order of then President Erich Honecker. In Romania, the armed forces, after at first participating in the shooting of unarmed demonstrators, refused to continue the slaughter and turned against their own commander in-chief and his personal security guards, who continued killing as many as they could. Clearly, the power of unarmed mass resistance, even against entrenched, ruthless governments, has been underestimated.

In addition, many critics, Thomas Lynch included, mistakenly assume that civilian resistance should be judged on the basis of whether it alone can do the work of an entire security policy. It never should be considered in isolation from other aids to security. When combined with significant steps toward demilitarization and the growth of world security and human rights institutions, it can be a vital asset in both maintaining peace and ensuring the responsiveness of governments that may be tempted to stray from accepted behavioral norms. It is not necessary, as Thomas Lynch seems to suggest, that "people power" must demonstrate that it can stop a missile attack in order for it to be worthy of consideration. After all, even the US armed forces in their entirety cannot repel a missile attack from a determined aggressor.

The limited attention given to the potential of mass protest as an instrument for the promotion and protection of important values suggests that many discussions of security focus too much on government-to-government relations. It may be important in the future to think more about pressures that people can exert directly on their own governments, whether to constrain military activities or to protect human rights. Transnational coalitions of organized citizens groups working together and in tandem with international peacekeeping and monitoring organizations, for example, could play a useful role in bridling the excesses of the existing balance of military power. In some ways, the efforts of citizens, encouraged by Mikhail Gorbachev's new thinking, have done more to set the stage for globally significant, positive changes in democratization, demilitarization, and reciprocity than have several decades of military alliances confronting one another.

Forming transnational alliances with non-violent demonstrators and building effective international support for the rights of indigenous peoples to press for reforms of their own governments often may serve US security better than external US military pressure. Surely military threats against communist governments in Eastern Europe could not have produced such unexpectedly positive results at such low cost as did primarily non-violent demonstrations in 1989. Therefore, to encourage steps in the direction of expanded political accountability, countries such as the United States could more pointedly attempt, through non-violent means such as conditional aid programs and promotion of international human rights covenants, to reform authoritarian regimes and to facilitate the growth of representative government. Although traditional ideas about sovereignty have restricted identities to territorial space, as R.B.J. Walker has pointed out, this confinement need not continue in the future. New security policies can support the growing feeling of individuals that they have responsibility for shaping their governments, a sense that now extends to every continent. The growth of people power, if it continues and remains essentially nonviolent, augurs well for the advancement of the other four principles proposed here.

Summary

More people become aware that there is no genuinely secure alternative to alternative security, then the policies proposed in this volume will surface in mainstream political thinking. As that happens, they will no longer be called alternative security policies.

This process is already well underway in some parts of Europe and among intellectual, religious, scientific, and artistic leaders in all countries. Alternative security ideas are spreading not because this is an era of optimism but because the old national security approach is, upon fair-minded reflection, thoroughly discredited. Military instruments and the militarily competitive balance-of-power system are unable to address many, if not all, of the most pressing, long range security problems of our era. Economic, environmental, human rights, and military questions can no longer be contained in national political arenas. Future success in constructing effective security policies rests on understanding that fragmented identities and stunted psychic development pose the underlying problems that must be overcome to generate the attitudinal value, and institutional changes that can enable our species to survive, and to survive with dignity.

Species identity and respect for all of creation are the keys to US and world security. With those keys in hand, the present generation is called to close the time-honored door of national foreign policy and to open the timeless door of world policy. We have reached the historic watershed when it is less important for one's own nation to possess the ability and the right to make war than it is to obtain influence over that ability and that right in the hands of others. To give up the former in order to obtain the latter would be a life-enhancing bargain. That bargain trades the war making function, now exercised by separate States operating under traditional national sovereignty, for a war-controlling function, to be exercised by many States operating with at least one sovereign function held in common.

Of course, there is nothing inevitable about progress toward the goals of alternative security. Indeed, one clings more to hope for rapid social learning than to confidence based on demonstrated ability to change resistant minds and institutions or to mobilize the hopeless, the indifferent, and the self-centered. Yet, by intentionally seeking to rise above the boundaries that separate nations and maintain the rich-poor divide, by cultivating respect for future generations and for all of nature, and by nurturing the growth and power of psychologically healthy and spiritually sensitive people to shape their own and each other's destinies, surely the development of a principled, peaceful and compassionate world policy will not exceed our grasp.

Bibliography

Abraham, Itty. The Making of the Indian Atomic Bomb: Science, Secrecy, andthe Postcolonial State. London: Zed Books, 1998.

Ahmed, Samina. Pakistan and the Bomb: Public Opinion and NuclearOptions. Notre Dame: University of Notre Dame Press, 1998.

Babbage, Ross, and Sandy Gordon, eds. India's Strategic Future. London: Macmillan, 1992.

Bailey, Kathleen C. Doomsday Weapons in the Hands of the Many. Urbana:University of Illinois Press, 1991.

Chari, P. R., Pervaiz Cheema, et al. Nuclear Non-Proliferation in Indiaand Pakistan: South Asian Perspectives. New Delhi: Manohar, 1996.

Chellany, Brahma. Nuclear Proliferation: The U. S.-Indian Conflict. NewDelhi: Orient Longman, 1993.

Dewitt. D. B., ed. Nuclear Proliferation and Global Security. London:Croom Helm, 1987.

Dunn, Lewis A. Controlling the Bomb: Nuclear Proliferation in the1980s. New Haven: Yale University Press, 1982.

Forsberg, Randall, Gregory Webb, and William Driscoll. NonproliferationPrimer: Preventing the Spread of Nuclear, Chemical and Biological Weapons. Cambridge, Mass.: MIT Press, 1995.

Frankel, Benjamin, ed. Opaque Nuclear Proliferation: Methodological andPolicy Implications. London: Frank Cass, 1991.

Gardner, Gary T. Nuclear Nonproliferation: A Primer. Boulder: LynneRienner Publishers, 1994.

Hagerty, Devin T. The Consequences of Nuclear Proliferation: Lessons fromSouth Asia. Cambridge, Mass.: MIT Press, 1998.

Joeck, Neil. Maintaining Nuclear Stability in South Asia. Adelphipaper no. 312. Oxford: Oxford University Press for the International Institute for Strategic Studies, 1997.

Kapur, Ashok. India's Nuclear Option: Atomic Diplomacy and DecisionMaking. New York: Praeger, 1976.

Lavoy, Peter. Learning to Live with the Bomb. Ithaca, N. Y.: CornellUniversity Press, 2000.

Mattoo, Amitabh, ed. India's Nuclear Deterrent: Pokhran II andBeyond. New Delhi: Har-Anand Publishers, 1998.

Meller, Eberhard, ed. Internationalization: An Alternative to NuclearProliferation? Cambridge, Mass.: Oelgeschlager, Gunn, and Hain, 1982.

Nanda, Ravi. Strategic Compulsions of Nuclear India. New Delhi:Lancer, 1998.

Palit, D. K., and P. K. S. Namboodiri. Pakistan's Islamic Bomb. New Delhi: Vikas Publishing House, 1979.

Quester, George H. The Politics of Nuclear Proliferation. Baltimore:Johns Hopkins University Press, 1973.

Reiss, Mitchell. Bridled Ambition: Why Countries Constrain TheirNuclear Capabilities. Washington, D. C.: Woodrow Wilson Center Press, 1995.

Sagan, Scott, and Kenneth Waltz. The Spread of Nuclear Weapons: ADebate. New York: W. W. Norton, 1998.

Tellis, Ashley. Changing Grand Strategies in South Asia. Rand Studiesin Public Policy. Cambridge: Cambridge University Press, 2000.

Waltz, Kenneth N. The Spread of Nuclear Weapons: More May Be Better. Adelphi paper no. 171. London: International Institute for Strategic Studies, 1981.

Yager, Joseph A. Nonproliferation and U. S. Foreign Policy. Washington, D. C.: Brookings Institution, 1980.

Zhang, Ming. China's Changing Nuclear Posture: Reactions to theSouth Asian Nuclear Tests. Washington D. C.: Carnegie Endowment for InternationalPeace, 1999.

Index

D

E